THE ICEPICK SURGEON

THE ICEPICK SURGEON

MURDER, FRAUD, SABOTAGE, PIRACY, AND OTHER DASTARDLY DEEDS PERPETRATED IN THE NAME OF SCIENCE

SAM KEAN

Little, Brown and Company
New York Boston London

Little, Brown and Company
Hachette Book Group
1290 Avenue of the Americas, New York, NY 10104
littlebrown.com

First Edition: July 2021

Little, Brown and Company is a division of Hachette Book Group, Inc. The Little, Brown name and logo are trademarks of Hachette Book Group, Inc.

ISBN 978-0-316-49650-6
Library of Congress Control Number: 2021935622

Printing 1, 2021

LSC-C

Printed in the United States of America

CONTENTS

Prologue: Cleopatra's Legacy 3
Introduction 6

1 Piracy: The Buccaneer Biologist 11
2 Slavery: The Corruption of the Flycatcher 33
3 Grave-Robbing: Jekyll & Hyde, Hunter & Knox 59
4 Murder: The Professor and the Janitor 79
5 Animal Cruelty: The War of the Currents 96
6 Sabotage: The Bone Wars 122
7 Oath-Breaking: Ethically Impossible 145
8 Ambition: Surgery of the Soul 174
9 Espionage: The Variety Act 201
10 Torture: The White Whale 229
11 Malpractice: Sex, Power, and Money 254
12 Fraud: Superwoman 283

Conclusion 300

Appendix: The Future Of Crime 307

Acknowledgments 323

Works Cited 325

Index 349

Most people say that it is the intellect which makes a great scientist. They are wrong: it is character.

—Albert Einstein

All I can say is, it's against the law to do many things, but the law winks when a reputable man wants to do a scientific experiment.

—Dr. Thomas Rivers

THE ICEPICK SURGEON

PROLOGUE

CLEOPATRA'S LEGACY

According to legend, the first unethical science experiment in history was designed by none other than Cleopatra.

At some point during her reign (51 to 30 BCE), a question arose among Egyptian scholars: when can you first tell whether babies are male or female in the womb? No one knew, so Cleopatra enlisted some maidservants in a fiendish plan.

This wasn't the queen's first foray into medical science. According to ancient sources — and modern historians back this up — Cleopatra took a lively interest in the work of her court physicians. She also invented a dubious cure for baldness — a paste of scorched mice and burnt horse's teeth, which was blended with bear grease, deer marrow, reed bark, and honey and massaged into the scalp "until it sprouted." More ominously, the Greek historian Plutarch reported that Cleopatra experimented on prisoners with poisons. She started with tinctures and chemicals — probably derived from plants — and graduated to venomous animals. (She even pitted different venomous beasts against each other in combat, fascinated to see who'd win.) This knowledge came in handy when Cleopatra ended her own life by letting an asp bite her breast, which she'd observed to be a relatively painless death.

As bad as poisoning prisoners seems, her experiment with fetuses surpassed that in depravity. We don't know the source of Cleopatra's obsession here—why she cared about the answer so much. But whenever one of her maidservants was sentenced to death (an apparently common occurrence), the queen ran her through the same procedure. First, in case she was already pregnant, she forced the maid to swallow one of the noxious substances she knew about, a "destructive serum" that purged the womb. With the slate now clean, Cleopatra had a manservant forcibly impregnate the maid. Finally, at some predetermined time later, she had the maid's belly torn open, and the fetus inside fished out. Accounts differ on the results, but Cleopatra could reportedly distinguish males from females by day 41 after conception—thus proving that sexual differentiation began early. All in all, she considered the experiment a success.

Now, the only historical mentions of this horror come from the Talmud, and on the face of it, the accounts are suspect. Cleopatra had scores of enemies who spread propaganda about her, and it's hard to think of a story that would demonize her more effectively than this. Furthermore, according to what doctors now know, the results don't make sense. Six weeks after conception, fetuses have eyes and a nose and little finger nubs, but they're only a half-inch long and don't have genitals—making it impossible to distinguish males from females. (Genitals form during week nine, when the fetus is two inches long.) So even laying aside the propaganda, it's doubtful whether Cleopatra performed this experiment.

Legend or not, however, many generations of people believed this story—which says something important. Cleopatra was powerful and hated, and the ghastly vividness of the tale gripped people's imaginations. We expect tyrants to do horrifying things. But even beyond that, something else about the account rang true. There was an archetype lurking there, something deep and scary that was recognizable even back then—a person who takes things too far and lets their obsessions get the best of them. What we now call a mad scientist.

The madness of a mad scientist is a peculiar one. They're not muttering gibberish or buttonholing you about loony conspiracies. To the contrary, they think quite logically. Here, Cleopatra experimented only on maidservants sentenced to death. If they were going to die anyway, she apparently reasoned, why not have them serve some useful purpose in the meantime? This decided, she made them take an abortifacient, to ensure that any prior pregnancies didn't confound her results. She then recorded the exact date of the rape-insemination, to nail down her answer to the day. If we judge this solely as an experiment, Cleopatra did everything right.

Judged by every other standard, of course, Cleopatra did nothing right. She grew so obsessed, so blinkered, that she abandoned all notions of decency and compassion—ignoring the gore and the shrieks of pain, pushing ahead no matter the human cost. No, what makes mad scientists mad isn't their lack of logic or reason or scientific acumen. It's that they do science *too well,* to the exclusion of their humanity.

Introduction

In our society, scientists are the good guys—usually. They're cool and clever, rational and clear-headed, calmly dissecting the world around us. But as the story of Cleopatra shows, sometimes obsession grips them. They turn things inside out and twist what's normally a noble pursuit into something dark. Under this spell, knowledge isn't everything—it's the only thing.

This book explores what pushes men and women to cross the line and commit crimes and misdeeds in the name of science. Each chapter is devoted to a different transgression—fraud, murder, sabotage, espionage, grave-robbing, and more: a comprehensive tour of the criminal arts. Admittedly, some of these stories are dastardly fun—who doesn't enjoy a good pirate yarn, or a juicy tale of revenge? Others, however, still leave us squirming centuries later. And while some of these incidents were splashed across the headlines of every tabloid in their day, many have been overlooked in history or receded with the fog of time, despite their sensational nature. This book resurrects these stories, and dissects what drove people to break the ultimate taboos.

These tales also have surprising things to say about how science works. We all know how discovery usually happens. Someone observes a curious event in nature, or gets a light-bulb idea into how some process or particle behaves. Then they test their hypothesis by running experiments or heading out into the field. If they're lucky, things go smoothly (ha). More often, frustrations pile up: experiments flop, funding gets yanked, fossilized colleagues refuse to accept new results. Finally, after much persistence, the evidence is too overwhelming to ignore, and the opposition thaws. The scientist returns from the intellectual wilderness, hailed as brilliant. The world at large benefits with a new medical treatment or high-tech material, or maybe even insight into where life comes from or the fate of the cosmos.

It takes a certain type of person to endure this gauntlet, someone patient and self-sacrificing. That's why our society has traditionally revered scientists as heroes. But science is more than a string of isolated eurekas. Like much of the rest of society, science has faced a moral reckoning recently, and understanding what good and evil look like in science—and the path from one to the other—is more vital than ever. Science has its own sins to answer for.

Even more surprising is the realization that unethical science is often ipso facto bad science—that morally dubious research is often scientifically dubious as well. At a glance, that might sound strange. After all, people often argue that knowledge is neither good nor bad; only human application makes it so. But science is also a communal activity—its results need to be checked, verified, and accepted by other people. Humans are baked right into the process, and as story after story in this book shows, science that ignores human concerns or tramples human rights consistently falls short of what it could be. At best, such work disrupts the scientific community and squanders time and energy on strife; at worst, it undermines the cultural and political freedoms necessary for science to take place. Harming and betraying people harms and betrays science in turn.

That's why these stories are of more than just academic or biographical interest. Only rarely do scientific villains emerge fully formed, like Athena from the head of Zeus. In most cases, morals erode slowly; people break bad step by painful step. By understanding what these scientists were doing, and why they thought themselves justified, we can spot the same dubious reasoning in modern research and maybe even prevent problems from arising. Indeed, dissecting depraved deeds provides an *opportunity* to learn how to short-circuit bad impulses and redirect people toward better ends.

Along those same lines, many stories here plumb the psychological motivations behind these warped deeds. What are scientifically minded criminals like? How do they differ from run-of-the-mill criminals? And how does their intelligence and advanced knowledge of the world aid and abet their wrongdoing? Chapter four, for instance, examines a sensational murder at Harvard, where a medical professor used his knowledge of anatomy to dispatch and dismember a university trustee. (He thereby became the second Harvard alum in history to be executed for a crime. In a later chapter, we'll meet the man who was nearly number three.) Many people just assume that intelligent folks are more enlightened and moral; if anything, the evidence runs the other way.

Finally, how do scientists justify their sins to themselves and others? Psychologists have actually identified several tricks that researchers use to rationalize their deeds and minimize their guilt—a primer on Why Good Scientists Do Bad Things. For one, scientists are more likely to trample ethical boundaries when they feel excessive pressure to reach their goals. Scientific scoundrels also employ euphemisms to mask their deeds, even to themselves. Or they perform a complex mental arithmetic, whereby the good they've done in the past somehow "cancels out" the harm they're causing now.

Scientists seem especially prone to tunnel vision. It's no secret that science rewards intense focus, and tunnel vision is a corollary of that focus. When immersed in their research, some people can't see beyond it, and they subsume everything in their lives to the pursuit of their goals, ethics included. In these cases, the morality or immorality of a research project might never occur to them. Chapter two recounts how many pioneering European scientists in the 1600s and 1700s—including giants like Isaac Newton and Carl Linnaeus—piggybacked on the transatlantic slave trade to gather facts and collect specimens from far-flung places. Yet few of them ever questioned their involvement in slavery, as long as the data kept flowing.

In still other cases, ethics get inverted. Compared to, say, politics, science seems pure. Just think of all the miseries science has liberated us from, all those life-saving medicines and labor-saving technologies. Scientists are justly proud of this record. But it's all too easy for some to fall into the trap that Science = Good, period. And in this worldview, anything that furthers scientific research must therefore be positive as well. Science becomes its own end, its own moral justification. Similarly, scientists with delusions of grandeur often fall for the means-end fallacy. They convince themselves that their research will usher in a scientific utopia, and that the bliss of that utopia will supersede, by many orders of magnitude, any suffering they're causing in the short term. Chapter five shows how Thomas Edison fell into this trap and tortured dogs and horses with electricity in order to prove the superiority of his preferred system of generating current. Even worse, chapter seven shows how research into eliminating sexually transmitted diseases has occasionally involved *giving* people syphilis or gonorrhea, in order to study them. In both cases, the reasoning was clear: just gotta crack a few more eggs. But when we sacrifice morals for scientific progress, we often end up with neither.

Beyond rationalizations, there's also the question of what makes scientific crimes unique. When regular people transgress, they do so for money or power or something grubby. Only scientists go rogue for data—to augment our understanding of the world. To be sure, many of the crimes detailed here are complex and have multiple motives; human beings are messy. Above all, however, these crimes spring from a Faustian drive for knowledge. For example, because of societal taboos against dissecting human bodies, many anatomists in the 1800s began paying "resurrectionists" to rob graves for them. Donning a black hat was the only way to get the knowledge they coveted. Some anatomists even robbed graves themselves, or accepted corpses from murderers. They grew so obsessed with their research that nothing else mattered to them, and their humanity got corrupted in the process.

These stories aren't just macabre antiques, either, something to dust off and scare students with—modern science is still reckoning with the fallout. Take the slavery-based research above. Many specimens collected via the slave trade became the nucleus of now-famous museums and remain on the shelves today. These museums wouldn't exist without slavery, which means that science and slavery are still intertwined centuries later. Or consider the experiments that Nazi doctors ran on prisoners during World War II. They tossed people into vats of ice water, for instance, to study hypothermia. It was barbarous work, and often crippled or killed the victims. But in some cases, it's the only real data we have, even today, on how to revive people in extremis. So what, ethically, should we do? Turn our heads, or use the data? Which outcome best honors the victims? Evil can roil a scientific field long after the perpetrators have died.

Beyond mining the past, this book contains some stories set in modern times, within the memories of people alive today. It also contains an appendix that looks to the fascinating future of crime. What dark deeds will scientists perpetrate in centuries to come? In some cases, like the crimes that will emerge when we colonize Mars and other planets, we can anticipate what will happen by looking at crimes on polar expeditions, where the bleak landscapes and sheer struggle to survive drove people mad. In other cases, there's really no precedent. What new crimes can we expect when we all have programmable robot companions in our homes, or when cheap, ubiquitous genetic engineering floods the world?

Overall, this book fuses the drama of scientific discovery with the illicit thrill of true-crime tales. The stories range from the dawn of science in the 1600s to the high-tech felonies of tomorrow, and they cover all corners of the globe. If we're honest with ourselves, we've all fallen into the rabbit hole of obsession before, or bent the rules in pursuit of something we coveted. But few of us have been as thoroughly corrupted as the rogues in *Icepick Surgeon*. We tend to think of science as progressive, a force for good in the world. And it usually is. Usually.

1

PIRACY: THE BUCCANEER BIOLOGIST

As the judge banged the gavel, William Dampier hung his head in disgrace. One of the most celebrated scientists of his age was now a convicted felon.

It was June 1702, and because this was a naval trial, the court had convened on the deck of a ship, exposed to the salty air. Everyone knew that most of the charges against Dampier had no chance of sticking. The murder claim was flimsy, and the accusations of being an incompetent navigator were laughable: he was the best navigator alive, a worldwide expert on winds, currents, and weather. But as the trial progressed, Dampier—who had long, limp hair and a hangdog look, with baggy eyes—sensed that the court was determined to punish him somehow, for something. And so it was: The judges found him guilty of thrashing his lieutenant with a cane on a recent voyage, and declared him "not a fit person to be employed as commander of any of Her Majesty's ships." He was fined three years' pay and dismissed from the navy.

Dampier staggered off the ship embittered and broken. How had he been reduced to this? He was the greatest naturalist of his day—so much so that Charles Darwin later counted himself a disciple. Dampier's sensational travelogues would also go on to influence *Robinson Crusoe* and *Gulliver's Travels*. No matter what he accomplished, however, William Dampier would always be guilty of one thing in the eyes of the Establishment. He was a brilliant scientist and navigator, no question. But for most of his adult life, he'd also been a pirate.

Given two things—his poverty and his obsession with biology—Dampier's descent into piracy was probably inevitable. He took to sea after being orphaned at age fourteen, visiting Java and Newfoundland before an unhappy stint in the navy. He eventually sailed to the Caribbean in April 1674 at age twenty-two. After some bouncing around, he settled in the Bay of Campeche in eastern Mexico and made his living cutting logwood, a thick tree whose inner pulp made a brilliant scarlet dye. Dampier later described his fellow loggers as a motley bunch, disposed to "carousing and firing off Guns 3 or 4 days together . . . [T]hey could never settle themselves under any Civil Government, but continued in their Wickedness." Though he probably caroused himself, Dampier also took long nature hikes in Campeche and was thrilled to see creatures he'd heard about before only in tall tales—porcupines and sloths, hummingbirds and armadillos. For someone who loved natural history, it was paradise.

His troubles started in June 1676, on one of those gorgeous early-summer days when it's almost a privilege to work outdoors. But while the other loggers basked in the sunshine, Dampier noticed that the wind was shifting directions oddly: it "whiffled about to the south, and back again to the east." Then the loggers noticed a mass of frigate birds overhead. These birds often accompanied ships from sea to shore, so most of the crew took this as a good omen; maybe supplies were coming. Dampier, however, frowned. The flock was

The pirate-biologist William Dampier was a big influence on Charles Darwin, as well as a rogue and a scallywag. Painting by Thomas Murray. (More pictures are available, for all chapters, at samkean .com/books/the-icepick-surgeon/extras/photos.)

Hitchcockian in its size and intensity, as if the birds were fleeing something. Most eerie of all was the local creek. Floods were a fact of life in Campeche; men often stepped directly from their beds into a pool of swamp water in the mornings. But that day the main creek started withdrawing mysteriously, as if sucked by a giant straw, until it was nearly dry in the middle.

Two days after these auguries, a bank of demon-black clouds rolled in and unleashed hell. None of the loggers had ever imagined a storm of such intensity. The rain stung them like hornets, blinding them, and the wind slapped down their huts one by one, until a single shelter remained. The men staggered through the mud toward it, shouting to be heard, and scrambled to shore up this last refuge with wooden posts and ropes lashed to tree stumps. It barely survived. Soaking and shivering, they then huddled inside for hours—and emerged into an alien world. The empty creek had more than filled back up, flooding the land around them. Trees were strewn everywhere, their roots forming impenetrable thickets. Dampier and a few other loggers managed to row their last surviving canoe over to the bay and found a shoal of dead fish floating upside down. Of the eight ships anchored in the bay hours earlier, all but one had been swept to sea. The loggers begged food from the crew of the surviving ship, "but found very cold entertainment," Dampier remembered. "For we could neither get Bread nor Punch, nor so much as a Dram of Rum."

Dampier's cinematic description of this storm was the first meteorologically detailed account of a hurricane, and it kicked off a lifelong preoccupation with wind and weather. More immediately, the storm rerouted the entire course of his life. All his logging equipment—axes, saws, machetes—had been washed away. He had no money and, without tools, no prospect of making any. As a result, he later wrote, "I was forced to range about to seek a subsistence." This was a euphemism. "Ranging about" meant becoming a buccaneer.

Buccaneers were a distinct class of pirates* then. Some pirates were so-called privateers, who had tacit permission from their home governments to harass enemy ships. English privateers usually focused on Spanish ships, and many an English home in the Caribbean was furnished with silks, pewters, and sleek carved chairs originally

* The late 1600s and early 1700s were the peak times of piracy for a reason. Several long wars in Europe had ended recently, meaning there were plenty of skilled sailors out there without jobs. While they could have joined the navy, many chafed at the harsh regulations then. There was also so much wealth zipping back and forth across the oceans, and so little way to police the vast seas, that it would have been shocking had pirates not flourished.

bound for Barcelona or Madrid. Privateering, then, was tolerated if not respectable. Buccaneers had no permission to raid anyone. They were simple criminals, and their home governments scorned them as much as their enemies did. The buccaneering crew Dampier joined stood even lower than most, because instead of raiding ships full of luxuries, they stormed pathetic little coastal camps, stealing from folks no better off than themselves.

We don't know what exactly Dampier did on these raids because he skimmed over most of the details in his journals, perhaps out of embarrassment. He also had a habit of getting distracted by natural history. Describing an assault on Vera Cruz, for instance, he dispatches with the death of a dozen companions in a few words, and skates right past the fact that the raid was a bust: the townsfolk had fled with their valuables at the first sign of pirates, leaving the town devoid of loot. Instead, Dampier highlights the dozens of caged parrots left behind, which he and others packed onboard his ship like legitimate treasure. They "were yellow and red," he gushed, "very coarsely mixed, and they would prate very prettily." No plunder, no matter—parrots were prize enough for him.

Dampier eventually returned to England in August 1678 and entered into a mysterious marriage with a woman named Judith, a lady-in-waiting to a duchess. Trying to go straight, he used her dowry to purchase some goods and sailed to the Caribbean again in January 1679 to trade them, promising his bride he'd return within a year. He broke that vow. A few months after arriving, he accompanied some sailors to Nicaragua on a trading trip, and the crew made a pit stop at a city in Jamaica that was a favorite haunt of lowlifes. Dampier later claimed to be shocked—shocked!—when the crew decided to throw their lot in with some pirates there and go buccaneering instead. In truth, some historians believe that Dampier knew full well that he'd meet pirates in Jamaica, and went there for the explicit purpose of returning to the high seas.

He did so for a few reasons. One, like every other person in history, Dampier longed to be rich, and there was always the chance that his band of buccaneers would stumble across a Spanish galleon groaning with doubloons and make their fortunes. But even deeper than that, Dampier couldn't shake his memories of Campeche—the rambles through the woods, the exotic flora and fauna, the whole days lost to nature. Pirating was the only means he had to recapture that feeling. To be sure, pirating was also a dirty business, rife with assaults and murders. Over the years Dampier saw priests being stabbed, prisoners tossed overboard, and native Indians picked off with rifles and tortured for intelligence. There's no reason to think Dampier stood aloof or was squeamish about his involvement. But Campeche had awakened a passion for natural history that was almost erotic in its intensity, and no matter how much he regretted buccaneering sometimes, his lust for new shores, new skies, new plants and animals, proved too strong. As he recalled, he was "well enough satisfied" wherever he ended up, "knowing that the farther we went, the more Knowledge and Experience I should get, which was the main Thing that I regarded."

Dampier joined the Jamaica pirates as a navigator, and the subsequent voyage was a rambling adventure, involving several different crews and ships; it's impossible to summarize neatly. They started off raiding cities in Panama, then sailed up to Virginia, where Dampier was arrested for unknown reasons; he refers to the incident only as some "troubles." Then it was over to the Pacific coast of South America, including the Galápagos.

Every so often the crew made a decent score: gems, bolts of silk, a stock of cinnamon or musk. Once they seized eight tons of marmalade. Much more commonly a galleon would give them the slip on the open seas and they'd slink off to try another port. Or they'd endure a long and futile siege of a coastal town, only to learn that the citizens had snuck out their treasures under the pirates' noses, leaving the buccaneers empty-handed.

Dampier and his crew nearly drowned during a fierce storm on their way to Indonesia. (Engraving by Caspar Luiken.)

Instead of making a fortune in South America, "we met with little . . . besides fatigues, hardships and loss," Dampier recalled. They sometimes had to drink "copperish or aluminous" water from "stinking holes of rocks," and they spent many a night outdoors with nothing but "the cold ground for our bedding and the spangled firmament for our covering." On one occasion, during a storm so violent that the men didn't want to risk raising the sails, Dampier and another comrade had to scramble up the rigging and hold their overcoats open to steer the ship bodily.

Hoping for better luck, the crew eventually struck out for Guam, a daunting trip of more than seven thousand blank miles. They staggered ashore fifty-one days later, nearly starved. Had things gone on much longer, Dampier later learned, the crew was plotting to murder and eat the captain and the officers, including him. (The captain took this news in remarkably good humor. He turned to his navigator and laughed, "Ah, Dampier, you would have but made them a poor meal!" "For I," Dampier explained, "was as lean as the captain was lusty and fleshy.") From Guam, the crew made excursions to China and Vietnam, and Dampier later became the first Englishman to set foot on Australia. In addition to studying flora and fauna in each spot, Dampier took advantage of his time on the open sea to study winds and currents, developing into a first-class navigator. Even those who despised Dampier had to admit that he had a near supernatural ability to judge where land lay beyond the horizon by reading the winds and currents.

Throughout these trips, Dampier changed ships several times, joining different crews. Sometimes the changes happened amicably, with no hard feelings; Dampier simply wanted to go somewhere new, and "No proposal for seeing any part of the world which I Had never seen before could possibly come amiss," he decided. In other cases, Dampier had to flee a despotic captain under harrowing circumstances, once by squeezing through a porthole in the dead of night. During such escapes, he typically took just one possession, the most valuable thing in the world to him—his field notes on natural history.

His final escape, in the South Pacific, proved particularly daunting. Longing to return home, he and a few comrades, including four Indonesian prisoners, slipped off to an island and secured a canoe there. On their first sally for freedom, they capsized, and Dampier spent three days drying his notes over a fire page by page. Their second attempt plunged them into a storm, and they spent the next six days on the open ocean thrashing their oars about and praying. "The Sea was already Roaring in a white Foam about us . . . and our little Ark in danger to be swallowed by every Wave," Dampier recalled. Worst of all, he hadn't been to confession in years, and scores of unnamed sins were weighing on his soul: "I made very sad Reflections . . . and lookt back with Horrour and Detestation, on Actions which before I disliked, but now I trembled at the remembrance of." Miraculously, they finally reached land at Sumatra, where Dampier collapsed onshore and spent six weeks recovering his strength. He eventually wound his way home on various ships and pulled into London in September 1691, a dozen years after he'd promised his new wife he'd be back in twelve months.

As a lady-in-waiting, Judith had her own life and had managed just fine without her rascally husband. But the pirate now had to support himself and make a living. And with few other options—you can't exactly put "buccaneer" on your resume—Dampier began organizing his field notes into a travelogue. That his journals had even survived the voyage was a miracle. They'd suffered water damage on several occasions, and he once had to stuff them into bamboo tubes for safekeeping. But the effort paid off.* *A New Voyage Round the World* finally appeared in 1697 and was a smash hit, with some of the liveliest passages on natural history and anthropology ever written.

After his stint in Sumatra, Dampier gave the first account in English of "ganga," or marijuana: "Some it keeps sleepy, some merry, some putting them into a laughing fit, and others it makes mad."

---

*We take writing for granted nowadays, but Dampier couldn't just grab a pen and start jotting. Every single time he found something worth recording, he had to dig out a quill from his chest below deck, hand-sharpen it with a knife, prepare some ink from powder and water, and find a place that wasn't too dark or damp or crowded with rowdy sailors—all just to get a few words down. The work didn't stop there, either. After writing, he had to sand the paper to sop up extra ink or risk it smearing, then stow it all away again and hope to god that vermin didn't devour it in the meantime. Writing is never an easy racket, but back then it was *work*.

He described a mass circumcision of twelve-year-olds in the Philippines, and how "they went straddling for a fortnight after." He also covered tattooing in Polynesia and foot-binding in China (which he denounced as a mere "stratagem" of the men to hobble women and keep them indoors). Upon hearing a local legend in the West Indies, he ate a dozen pickle pears at once and delighted to see that they really did turn his urine red. Nearly a thousand citations in the *Oxford English Dictionary* trace back to his writings, and he introduced dozens of words into English, including *banana, posse, smugglers, tortilla, avocado, cashews,* and *chopsticks.*

There was plenty of science, too. Even today Dampier remains unsurpassed as a pure observer of nature. Compared to his, other accounts of flora and fauna seem lifeless—like a glass-eyed taxidermy lion compared to the real, leaping, roaring beast. Part of the vividness sprang from Dampier's use of all five senses, including taste. The man never met an animal he didn't eat. Flamingo tongues, he reported, have "a large knob of fat at the root, which is an excellent bit; a Dish of Flamingos Tongues being fit for a Prince's Table." He cooked up manatee veal and iguana soup and turtle-oil dumplings, along with dozens of other outré recipes. And if all that got your mouth watering, well, Dampier could snatch your appetite away just as quickly. In one revolting section he popped a worm cyst on his leg and drew out the slimy sucker inch by agonizing inch. He also detailed—my apologies in advance—one of the most epic bouts of diarrhea ever recorded in the annals of English literature. His attack started after seeking treatment for a fever, when he was persuaded to take a local "physic" to purge his bowels. Bad idea. He had the trots off and on for a year, and sometimes endured thirty bowel movements in a single sitting, till he was more or less dry-heaving out his arse. No one ever said field work was glamorous.

The quintessential Dampier tale involved an alligator attack. He opened the story with a passage noting the differences between alligators and crocodiles. In an era when most scholars still lumped whales with fishes, the ability to make such fine distinctions was impressive, and wouldn't look out of place in a herpetology text today. Then things swerved. Without any transition, Dampier launched into the tale of a nighttime hunting expedition back in Campeche. On it, an Irishman named Daniel tripped over an alligator, which whipped around and clamped down on his leg. He screamed for help. But his companions, Dampier noted, "supposing that he was fallen into the clutches of some Spaniards," abandoned him, leaving him alone in the dark with a gator gnawing on him.

Amazingly, Daniel kept his cool and came up with a plan. Unlike mammals, reptiles lack lips and can't chew. They gulp food down in big bites instead and have to open their mouths to pull prey farther inside. So when his tormentor opened its jaw again, Daniel shot forward and jammed his rifle inside in place of his leg. Fooled, the gator pulled the gun away to devour it, and while he did so, Daniel scrambled back.

Surging on adrenaline, he dragged himself up a tree and renewed his cries for help. His comrades, realizing there were no Spaniards afoot, returned with firebrands to drive the gator off. Afterward, Dampier reported, Daniel "was in a deplorable condition, and not able to stand on his Feet, his Knee was so torn with the Alligator[']s Teeth. His Gun was found the next day . . . with two large Holes made in the But[t]-end of it, one on each side, near an Inch deep." All in all, the tale was classic Dampier—learned, meticulous, and hair-raising all at once.

Some historians credit *A New Voyage* with launching the entire genre of travel writing, and after its publication, Dampier received an invitation to lecture at the prestigious Royal Society in London, the world's premier scientific club. Not bad for a buccaneer. He also dined with several eminent statesmen, including the diarist Samuel Pepys. The bigwigs wanted to talk natural history, of course, but some of them no doubt felt a frisson of pleasure in knowing they had a real-life pirate at their table.

With the public clamoring for more, Dampier published a sequel to *A New Voyage* in 1699. It included his famous essay "Discourse on

Winds," which later captains like James Cook and Horatio Nelson considered the best practical guide to sailing they'd ever read. The essay also greatly advanced the scientific study of winds and currents. Two of Dampier's contemporaries, Isaac Newton and Edmond Halley (of comet fame), had recently published treatises on the origins of tides and rainstorms, respectively. Dampier's essay then nailed down where winds and currents come from. In one sweep, then, these three scientists solved several age-old mysteries about the sea and the cyclical movement of water around the globe. We normally wouldn't include a pirate in the same company as Halley or Sir Isaac, but Dampier was every bit their equal in this field.

Oddly, though, Dampier wasn't in England to see his second book published. He'd actually made very little money from the first one, in part because copyright laws didn't exist then and most of the book's profits had been scooped up, ironically enough, by literary pirates. Dampier still needed to earn a living somehow. Moreover, he desperately wanted to leave piracy behind and reinvent himself as a respectable scientist. So the president of the Royal Society introduced Dampier to the First Lord of the Admiralty, who offered him the chance to captain his own ship, the *Roebuck,* and lead an expedition to New Holland (modern Australia). Despite any reservations he might have had about rejoining the navy, Dampier accepted. Part of his mission was to scout out commercial opportunities down under. But the main goal was scientific; this was the first explicitly scientific voyage in history. It was the most noble idea anyone had ever heard of. And with Dampier in charge, it was a disaster from the get-go.

Dampier had an air of Thoreau about him: a curmudgeon who delighted in nature but grumbled about his fellow human beings. He also had an arrogant streak. The pirate ships Dampier cut his teeth on often had a surprisingly democratic ethos. Some even had rudimentary health insurance, with a sliding compensation scale for lost eyes and limbs.* Dampier, though, was eager to distance himself from his past, and abandoned all such camaraderie aboard the *Roebuck*. He decided that he was smarter than everyone else in all matters, scientific and otherwise, and he lacked the charm and political skills to quell the unrest that resulted.

Especially when dealing with officers. Dampier's second-in-command, a naval lieutenant named George Fisher, despised Dampier as pirate scum. He swore up and down that Dampier was plotting to commandeer the *Roebuck* and go privateering as soon as they hit open water. The *Roebuck* set sail in January 1699, and even before it reached its first stop (the Canary Islands, to stock up on brandy and wine) Dampier and Fisher were quarreling. As one witness reported, in that blunt way of sailors, Fisher "gave the captain very reproachful words and bade him kiss his arse and said he did not care a turd for him."

Those tensions erupted into violence in mid-March. Like so many troubles in life, it started with a keg of beer. It was a nautical tradition to tap a keg whenever a ship first crossed the equator, to let the men blow off some steam in the torrid weather. Dampier's crew ran through their keg a little quickly, though, and complained that their throats were still parched. They begged Fisher to let them tap a second. Instead of consulting Dampier, as navy regulations required, Fisher gave assent alone.

This was hardly mutiny. But Dampier was already on edge: Rumors were swirling that Fisher planned to heave him overboard to

---

*Pirates got 600 pieces of eight for losing a right arm, 500 for losing a left arm or right leg, 400 for losing the left leg, and 100 for losing an eye or finger. This was all drawn up in official documents, since a surprisingly high percentage of pirates could read (around three-fourths), largely because they need to understand charts. Pirates also held elections on where to sail next for plunder (a simple majority ruled), and their mealtimes were surprisingly democratic as well. They shared all food equally and, unlike in the snobbish navy, the officers couldn't poach the choice bits.

feed the sharks. And this deliberate undermining of his authority snapped the last frayed band of restraint he had left. Upon seeing the second keg, he grabbed his cane, found the cooper who'd tapped it, and cracked him upside the head. Then he wheeled on Fisher and demanded to know why he'd allowed this. Before Fisher could answer, Dampier clubbed him, and proceeded to beat him bloody. He then clapped Fisher in leg irons and confined him to a locked cabin for two weeks. Fisher couldn't leave even to use the head, and had to stew in his own filth. When the ship reached Bahia, on the Brazilian coast, Dampier had his lieutenant arrested and jailed without food.

If Dampier thought he'd won this power struggle, however, he miscalculated. The moment his cell door clinked shut, Fisher climbed up to his window and began yelling at passersby in the street, railing about his imprisonment and slandering Dampier left, right, and sideways. He later composed letters to authorities in England to expose the pirate-scientist as a tyrant. Fisher's every waking thought was to destroy Dampier.

Dampier, in contrast, dealt with the matter by burying his head in natural history. While Fisher plotted, Dampier disappeared into the bush around Bahia, taking notes on indigo and coconuts and tropical birds. One observation in particular on this trek stands out for its historical importance. After observing some flocks of "long-legg'd fowls" at different sites, Dampier realized that, while each flock was distinct, no one group was distinct enough to count as its own species. There was a continuum of variation. So he coined a new word, "sub-species," to describe this state. That might seem like a minor insight, but Dampier was groping toward an idea—about variation in nature and the relationships among species—that his admirer Charles Darwin would later run with in *On the Origin of Species*.

The Catholic Inquisition in Brazil finally put a stop to Dampier's hikes. They didn't relish the thought of a Protestant pirate wandering around and taking notes on everything, and there were rumors afoot that the church planned to arrest or even poison him. Perhaps fearing he'd end up in chains next to his second-in-command,

Dampier made haste to sail. He also arranged to ship Fisher back to England, to what he no doubt assumed would be a humiliating trial for insubordination. Dampier was only half right. There would be a trial, and humiliations aplenty, just not for Fisher.

With Fisher absent, tensions cooled onboard the *Roebuck,* and by mid-August the crew reached western Australia, landing on the gleaming white beaches of Shark Bay. They spent the next few weeks observing dingoes, sea serpents, humpback whales, and more—a brilliant start to their scientific campaign.

Their luck didn't hold. Western Australia is as bleak as it is arid, and despite scouring the coast, the *Roebuck* found zero sources of fresh water. The sailors were soon desperately thirsty, so they tried approaching some Aborigines, who they assumed had tricks for finding water. (The Aborigines did, including tracking birds and frogs and hacking at tree roots.) But whenever the sailors came near, the natives scattered. So Dampier came up with a desperate plan. After creeping ashore, he and two companions hid behind a sand dune to ambush the natives. The plan was to kidnap one and force him to lead them to a spring. When the Englishmen jumped out, the Aborigines once again ran, and the Englishmen gave chase—not realizing they were falling into a trap. As soon as Dampier and company were exposed on open ground, the Australians wheeled and attacked with spears. One of Dampier's men was slashed in the face, and Dampier himself almost got impaled. When warning shots failed to drive the natives back, Dampier took aim and wounded one with his pistol. It's a rare moment in his books where he admits to committing violence.*

*There's no doubt Dampier would flunk any modern test of enlightenment; like all people then, he had his prejudices. But his biographer referred to him as a "humane man in a not very humane time," which sums things up pretty well. In fact, if you take the time to actually read his works, what stands out most is his tolerance of foreign cultures. Whenever he saw a (to him) strange custom or

Realizing they'd never find water now, Dampier's crew slunk away from Australia in disgrace, and things only got worse from there. After Australia, Dampier tried to salvage the voyage by exploring New Guinea and collecting specimens. But the English navy hadn't exactly handed him its trustiest ship in the *Roebuck*. Its hull leaked and was infested with worms, and it soon got so creaky that Dampier had to turn tail and make a run for England. The ship never arrived. On the shores of Ascension Island in the southern Atlantic, the *Roebuck* sprung a fatal leak. Fearing he'd be blamed for its demise, Dampier tried plugging the hole with anything he could think of, including a side of beef and his personal pajamas. Both stopgaps failed. The crew abandoned ship at Ascension and Dampier lost virtually every specimen he'd gathered. The men spent five weeks watching other ships sail blithely by in the distance until a flotilla finally pulled in and rescued them.

Returning to London without specimens, not to mention the ship, was bad enough. But when Dampier arrived in August 1701, he found that George Fisher had been poisoning English society against him—damning the ex-pirate with such lusty broadsides that the admiralty felt obliged to court-martial Dampier and try him aboard a ship.

Dampier defended himself as best he could, marshaling witnesses who swore that Fisher was plotting mutiny. Dampier also fought dirty, accusing Fisher—no one knows how truthfully—of sodomizing two young cabin boys on their voyage. (Pirates tolerated

---

rite, he always refrained from judgment and strived hard to understand it. He was also much harsher in judging his own countrymen. He rolled his eyes, for instance, when other officers refused to trust a mixed-race female prisoner as a guide, simply because of who she was. So while hardly woke by modern standards, the buccaneer biologist seems remarkably tolerant for his time, and he recognized that Europeans usually brought violence on themselves: "I am of the Opinion that there are no People in the World so barbarous as to kill a single Person that falls accidentally into their Hands, or comes to live among them, except they have before been injured, by some outrage, or violence committed against them."

homosexuality to some degree; the navy did not.) For his part, Fisher harped on Dampier's character, denouncing him as a poltroon and a scoundrel. He also ginned up charges that Dampier had murdered a querulous crew member by locking him into a cabin for a spell, even though the man died ten months after the punishment ended. To their credit, the judges dismissed that and other charges, including one of negligence for letting the *Roebuck* sink. But they could not abide the caning of Fisher, a fellow officer, and they found Dampier guilty of "very hard and cruel usage" of his second-in-command. As punishment, they banned him from commanding any English ships and fined him three years' salary.

William Dampier had tried to go respectable, and it had profited him nothing. He was as penniless as ever, and was now a pariah in government circles. He had only one option left: the 49-year-old naturalist would have to return to piracy.

Dampier's life and times might seem remote to us, but the ethical issues he raises are still relevant today. For one thing, scientific piracy didn't end in the 1700s. Moreover, the kind of fieldwork he did is, in some ways, even more dangerous now than several centuries ago.

Countless naturalists over the years have died with their boots on. Most succumbed to malaria, yellow fever, or another disease, but there are enough snake bites, stampedes, puma maulings, mudslides, and accidental poisonings to fill a whole volume. Scientists have been murdered, too. In 1942, Ernest Gibbins, a British biologist studying blood-borne diseases in Uganda, was ambushed in his car and stabbed to death by local warriors who were convinced he was stealing their blood for "white man's witchcraft." A police officer said his body was "as full of spears as a bloody porcupine." Since then, a rise in tribal wars and ethnic conflicts in the twentieth century, exacerbated by global arms trafficking, has only increased the danger of fieldwork in many places. Dampier and his contemporaries suffered gravely at times, but he never had to worry about being kidnapped and held for a $6 million ransom by an armed militia, as happened to a rice scientist in Colombia in the 1990s. For these reasons, many research institutes are far less tolerant nowadays of the haphazard, let's-just-wing-it fieldwork of yesteryear.

As for scientific piracy, its nature has changed since Dampier's day. Again, Dampier became a pirate largely to feed his scientific obsessions; he had no other means to visit distant lands. In contrast, with later scientists, the very nature of their work was criminal in that it involved the theft of natural resources—so-called biopiracy.

One highly coveted good during colonial times was quinine, a drug derived from the cinnamon-colored bark of the cinchona tree. When ground into powder and drunk with water, quinine helps combat malaria, the deadliest disease in human history. (According to some estimates, mosquito-borne diseases have killed a full one half of all 108 billion human beings who've ever lived, and malaria represents the biggest chunk of that carnage.) Unfortunately, while malaria was a worldwide scourge—killing people in Africa and India, Italy and Southeast Asia—cinchona trees grew only in South America. So European nations began sending botanists undercover into South America to steal cinchona seeds. It proved a fool's errand. The most valuable, quinine-rich species lived on stupidly steep slopes in the Andes that were shrouded in mist three-fourths of the year. As a result, every single smuggler failed, and several perished in the attempt.

The man who finally succeeded was a Bolivian Indian named Manuel Incra Mamani. Very little is known about Mamani. Stories that he descended from an Incan king are almost certainly bogus, though he might have come from a line of medicine men who prized botanical knowledge. Regardless, he could tramp through the Amazon for weeks at a time, fueled by little more than coca leaves, and he had an uncanny ability to scan the endless green canopy of the forest and pick out a tiny wisp of scarlet—the signature color of cinchona leaves. After harvesting a few sacks of seeds in 1865, he tramped a thousand miles, on foot, over the freezing Andes highlands, and delivered them to the Englishman who'd commissioned him. For this deed, he got $500, two mules, four donkeys, and a new gun. He was also sentenced to death in absentia for betraying his country. The greedy Englishman later sent him back into the jungle for more seeds, at which point Mamani got caught and charged with smuggling. He was thrown into jail, denied food and water, and beaten savagely. He was released two weeks later, so crippled he couldn't stand upright. His donkeys were taken from him, and he died within a few days.

Historians still debate whether Mamani's crime was justified. On the one hand, Peru and Ecuador had been hording an essential medicine and charging wildly inflated prices—profiteering on death. Moreover, they'd been husbanding the trees so poorly that cinchona was on the verge of extinction by the mid-1800s. After Mamani, several European nations established cinchona plantations in Asia with the smuggled seeds, thereby saving millions of lives worldwide.* (Incidentally, British officers in India consumed the bark as bitter tonic water, which they mixed with booze to make it go down more smoothly. Thus was born the gin and tonic.) On the other hand, the plantations in Asia undermined and eventually wiped out the native cinchona industry in South America, impoverishing people there. And given the value of cinchona as a medicine, one historian has called the theft, only slightly hyperbolically, "the biggest robbery in history." It was colonialism at its most exploitative. It also saved countless lives in Africa and Asia.

---

*If you like Nazi stories—and let's be honest, every story is better with some Nazi villains—then I encourage you to check out my podcast, called The Disappearing Spoon, for a heckuva tale. It explains how a few crooked Nazis probably saved more American lives than anyone else in all of World War II by supplying us with quinine at a desperate time. Overall, the podcast contains all new stories, ones that don't appear in my books. You can subscribe via iTunes, Stitcher, or any other platform, or visit my website at samkean.com/podcast.

Other biopiracy seems harder to justify. One key ingredient in industrialization was rubber, which was derived from the sap of certain trees native to the Amazon. Without rubber tires, cars and bicycles wouldn't exist, and rubber tubes and seals made modern chemistry and medicine possible. Nor would we have electricity without rubber insulation for wires. But rubber remained a niche good until British explorer Henry Wickham broke Brazil's monopoly on it in 1876 by smuggling out 70,000 rubber seeds, which were used to set up more plantations in Asia. The world at large benefitted, no question, but stealing seeds to make consumer goods seems less ethical than stealing seeds to make medicine. Other cases of smuggling seem even less moral. Consider the Scottish botanist in China in the 1840s who dressed up in local garb, shaved the front half of his head, pulled his remaining hair into a ponytail, infiltrated a state-run plantation, and stole 20,000 prize tea plants for transport to India. You'd be hard-pressed to make a humanitarian case for Earl Grey.

Biopiracy continues in modern times. Billionaires in China pay fortunes to poachers for rhinoceros horns and other supposed priapics. Pharmaceutical companies develop blockbuster drugs from viper venom and periwinkle plants and other tropical resources, and only rarely does the money trickle back down to the indigenous peoples who, in some cases, first discovered their medicinal properties. It's not all the mega-rich, either: everyday people around the world support an extensive black market of exotic flowers and pets. Even if the offenders aren't hunting down doubloons and pieces-of-eight anymore, the spirit of Dampier-era piracy lives on.

In 1703, William Dampier finally caught a break. A new war with Spain and France had broken out, and England needed privateers to harass the enemy. So despite his being barred from commanding her ships, Queen Anne summoned the 51-year-old pirate for an audience. Like the basest courtier, Dampier kissed the royal hand and kissed the royal ass, and soon received a commission to captain the *St. George*.

Alas, the voyage of *St. George* was another messy, mutinous affair. Dampier's men accused him of taking bribes (e.g., silver dinnerware) from the captains of the foreign ships he'd captured. In exchange, Dampier would conduct only a superficial search of their holds, and let them sail away with most of their treasures intact. Rumor also held that Dampier was drinking heavily, although it's hard to blame him. He spent all day every day tacking back and forth, back and forth, scouring the horizon for distant ships. It was dreadfully boring, and unlike in his buccaneering days, he couldn't just chuck his duty and sail for some distant port. He had responsibilities now, and his inability to indulge his scientific curiosity left him miserable. (Modern research shows that IQ correlates strongly with alcohol abuse, and it stands to reason that people would drink more when they feel intellectually thwarted.) When the voyage ended in 1707, Dampier's reputation as a captain was in shambles, and he never again commanded another ship.

However poor a captain, though, Dampier was still a brilliant navigator, and a few years later he joined another privateering cruise that made literary history. On an excursion in the Pacific Ocean the crew began running low on water and drooping from scurvy, so Dampier steered them toward the nearest land, the Juan Fernández Islands off Chile. While approaching, they were astonished to see a hairy bipedal beast onshore, waving its arms. It was a marooned sailor named Alexander Selkirk. He was clad in goatskins and looked, as one witness recalled, "wilder than the first owners of them." For four years, four months, and four days Selkirk had eked out a living on the island—snagging goats, gnawing wild cabbages, making knives and fishhooks from barrels that washed ashore. His feet were as leathery as iguana hides, and after four years of isolation, his voice was so hoarse he could barely speak. Dampier's crew rescued him and brought him back to England in triumph. His story would soon inspire Daniel Defoe to write *Robinson Crusoe*.

Dafoe was hardly alone in mining Dampier's life for inspiration. Jonathan Swift poached his stories for *Gulliver's Travels,* and Samuel Taylor Coleridge did the same for "The Rime of the Ancient Mariner." Dampier's most influential fan, Charles Darwin, even brought Dampier's books along on his formative *Beagle* voyage in the 1830s. Darwin chuckled over his pirate predecessor's naughty deeds, calling him "Old Dampier" in his notes. More importantly, Darwin studied Dampier's descriptions of species and subspecies and pored over his accounts of places like the Galápagos, effectively using Dampier as a guide. Darwin might never have become Darwin without the old pirate.

But while adventure writers and scientists have always forgiven Dampier, latter-day George Fishers have found him harder to abide. When Dampier's hometown in England discussed putting up a plaque to honor him in the early 1900s, one god-fearing fellow stood up and denounced him as "a pirate ruffian that ought to have been hung [sic]." Critics nowadays go even further. They contend that Dampier's science, however groundbreaking, merely blazed a trail for colonialism and was therefore a crime against humanity.

The thing is, both sides have a point. Dampier was lowdown and brilliant, inspiring and a blackguard. His work advanced nearly every scientific field that existed then—navigation, zoology, botany, meteorology—and he did despicable things in the meantime. As one biographer noted, "Defoe, Swift, and all of the others owed much more to Dampier than a single model. It can in truth be said that they owed the entire spirit of a new age to this one man."

Alas, this new age would have atrocities of its own to reckon with—especially slavery. At a glance, science and slavery might not seem to have much to do with each other. But both were fundamental forces in shaping the modern world, and historians have started to recognize that they shaped each other in disturbing ways as well.

2

SLAVERY: THE CORRUPTION OF THE FLYCATCHER

When the Englishman Henry Smeathman set sail for Sierra Leone in October 1771, he had every reason to think his expedition would be a triumph. At twenty-nine, he was the perfect age for a naturalist—old enough to have experience, young enough for adventure. And given all the outlandish specimens pouring into Europe then from across the globe—orangutans and goliath beetles, Venus flytraps and "flying cat-monkeys" (i.e., flying squirrels)—he had high hopes of making his own grand discoveries in Africa.

Wasting no time, Smeathman and his assistant started collecting on the voyage down, breaking out their nets on deck and snagging butterflies and locusts blown out to sea. True, most of the specimens were soon devoured by the ants and cockroaches on their filthy ship, the aptly named *Fly*. Yet the ever-cheerful Smeathman devised a workaround in no time. After placing his specimens atop a tapped keg of rum, he found that the fumes discouraged vermin. He jotted this down in his journal as "a useful tip for naturalists."

The *Fly* finally reached Africa on December 13, dropping anchor at the Îsles de Los, an offshore trading post for ivory and lumber;

Smeathman described it as "little mountainous islands covered with trees & shrubs." It should have been a satisfying moment: the end of the cramped voyage, the commencement of his scientific work. But Smeathman tensed as he descended the gangplank. For the Îsles were more than just a marketplace for luxury goods. They were also a place of chains and whips—the epicenter of the Atlantic slave trade.

Before setting out, Smeathman had known that slavery would form the backdrop for his journey. He was a determined foe of slavery, and when pitching his trip to his sponsors, he'd vowed to tell the truth about "those little-known and much misrepresented people, the Negroes." But even this determination couldn't prepare him for the shock of seeing slavery for himself.

Upon arriving at the Îsles, Smeathman and his fellow passengers toured a slave ship, the *Africa*. The sensory assault began even before entering, Smeathman wrote: "Our ears were struck at some distance with a confused noise of human voices & the clanking of chains, which . . . affects a sensible being with inexpressible horror." Onboard, the male slaves had been stripped naked, supposedly for health reasons, while the women wore just loincloths. Smeathman was especially distressed to see two women breastfeeding their infants amid the chaos; he said he'd never seen sorrow "more strongly marked in the human face." The rest of his group kept strolling and chatting, as if on a garden tour, but Smeathman kept glancing back at the mothers. "They would undoubtedly have shed tears," he added, "if they had had hopes of compassion, or their nature had not already been exhausted. I was absorbed in a thousand melancholy reflections and bore a very small part in the conversation."

He also met the *Africa*'s captain, John Tittle. Tittle was vicious even by the standards of slave traders, a trait that led to his grisly demise a few years later: After dropping his hat into a harbor one day, Tittle ordered a small Black boy in his employ to dive in and retrieve it. The boy refused; he feared sharks and couldn't swim. Tittle threw him in anyway and he drowned. Had this been a slave boy, no one would have dared confront him. But Tittle had murdered the son of a local chief, who demanded recompense in the form of rum. Tittle responded by sending several barrels—filled not with rum, but with "emptyings from the tubs of his slaves," possibly including excrement. The enraged chief hunted the captain down and clapped him in irons. He then starved Tittle and tortured him to death while local villagers—who were equally sick of Tittle's crap—gathered around and howled in delight.

Despite Tittle's sadistic reputation (or perhaps because of it) slave companies happily entrusted him with the lives of their "cargo." The *Africa* was designed to hold 350 slaves, but not long after Smeathman's visit, Tittle packed 466 souls into his hold and sailed for the Caribbean. Eighty-six men, women, and children died en route.

To Smeathman's relief, his party soon quit the Îsles and sailed for Bunce Island near the African mainland. But he couldn't escape slavery there, either. Bunce was an odd, almost schizoid place, once described as half slave port and half "country estate," complete with a two-hole golf course. The fort there boasted cannons and sixteen-foot walls, a defense against Dampier-like pirate raids.

The slave-traders on Bunce, always eager for news from home, buttonholed Smeathman and peppered him with questions. If they were dressed like typical slavers, they wore checked shirts and black handkerchiefs tied around their necks or waists. Smeathman chatted happily about England for a few minutes, but the conversation soured for him as soon as they asked his reasons for visiting Africa. When he revealed his interest in natural history, they laughed in his face. As one slaver said, "The longer one lives the more one learns! To think now of anybody coming two or three thousand miles to catch butterflies and gather weeds." Some began openly mocking him.

Smeathman snorted and turned his back on them—consoling himself with the thought that, while they'd come to Africa to sell women and children into bondage, he'd come as a scientist, to advance knowledge and improve the lot of humankind. He'd have nothing to do with these barbarians.

That superiority would prove hard to maintain, however. In coming to Africa, the young naturalist had been sailing not only toward something but away from something—the old Henry Smeathman. The old Smeathman was a pauper, a failed striver, someone he wanted to abandon and bury back in England. This expedition marked the debut of the new Smeathman, the gentleman-naturalist. Much like with William Dampier, he felt science was his best shot at creating a better life for himself. In rejecting the slavers, then, he was rejecting both their morals and their lowly station in life.

Ultimately, though, Smeathman's ambition to remake himself as a scientist would prove stronger than his morals. Despite his opposition, slavery so dominated the economy of Sierra Leone that he soon found himself trading with slavers for supplies and equipment. Before long, he was doing even worse things. Predictably, too, the more entangled he got, the more he felt a need to defend his trading partners—and by extension, himself. It was a textbook psychological defense: *I'm a good person and would never associate with bad people. Therefore, the people I* am *associating with can't be that terrible.* But once he started down this road of rationalization, it proved slipperier than he ever imagined.

Amid the vast atrocities of the slave trade, the corruption of a single "flycatcher" like Smeathman hardly stands out as tragic. (This should go without saying, but given how charged this topic is, it's worth being explicit: the Africans were the victims here, not the white European.) Still, Smeathman's life is worth examining, because it sheds light on an aspect of early science that most historians overlook—how intertwined science and slavery were. Moreover, Smeathman's story reveals just how easily slavery could corrode the morals of even sincere, well-meaning people. Far from being a backdrop, the slave trade would come to dominate his time in Sierra Leone. Bit by bit, compromise by compromise, it would twist his ethics inside out.

Slavery is as old as civilization, but the transatlantic slave trade between the 1500s and 1800s was exceptionally brutal. Estimates vary, but at least ten million Africans were enslaved during wars and raids, with roughly half dying on the march to local ports or on voyages across the ocean. And statistics alone can't capture the cruelty of slave ships. Men, women, and children were chained up in holds so hot and filthy that the stench of bodies often caused people to vomit upon entering. Toddlers sometimes stumbled into the "necessary tubs" of human waste and drowned. Diseases ran rampant, and the sick were often tossed overboard to spare the others. (Sharks in fact followed slave ships sometimes, for easy meals.) Slaves might also be tossed to the sharks for disobedience—or subjected to worse punishments. After one failed slave revolt onboard a ship in the 1720s, the captain forced two of the instigators to kill a third and eat his heart and liver.

So why did scientists align themselves with this horror? Access. European governments did sponsor scientific expeditions now and again, but the far majority of ships visiting Africa and the Americas then were private vessels engaged in the so-called triangular trade—a tripart exchange that sent guns and manufactured goods from Europe to Africa; slaves to the Americas; and dyes, drugs, and sugar back to Europe. Outside that trade, travel options to Africa and the Americas were nil. Field scientists determined to visit those lands therefore hitched rides on slave ships. Upon arrival, they also depended on slave merchants for food, supplies, local transport, and mail.

Naturalists who stayed behind in Europe* took advantage of the slave trade as well. In many cases they deputized the crews of

---

* Stay-at-home collectors were sometimes scorned as "armchair naturalists" because their ignorance of real, living plants and animals could lead them to absurd conclusions. For instance, when one species of Papuan songbird first reached Europe in the mid-1700s, collectors named it the "bird of paradise," both for its lovely plumage and the fact that it lacked feet—an indication, they decided, that the bird never needed to land. It spent its entire life aloft instead, swooping through the heavens. In reality, the natives who'd captured the bird simply used

slave ships to collect on their behalf—especially ship surgeons,* who had scientific backgrounds and enjoyed plenty of free time on shore while their crewmates sold slaves and bought provisions. The specimens collected—ostrich eggs, snakes, butterflies, nests, sloths, shells, armadillos—were then transported back to Europe on slave ships, before eventually finding their way into research institutes or private collections. Carl Linnaeus, the father of taxonomy and one of the most influential biologists in history, drew upon such collections when putting together his monumental *Systema Naturae* in 1735—the book that introduced the genus-species naming system of *Tyrannosaurus rex* and *Homo sapiens* that we still use today. Overall, these collections were the "big science" of their time—centralized repositories that were crucial for research projects. And they were all built on the infrastructure and economics of slavery.

Henry Smeathman, however, thought he could sidestep this moral morass. No known portrait of Smeathman exists today, and the one surviving description of him is enigmatic: "tall, thin, lively, and very interesting, but not handsome." As a boy he'd loved collecting shells and insects, but his formal education was cut short when his tutor, a curate, killed himself. After that he tried his hand at making cabinets, upholstering furniture, selling insurance, distributing liquor,

---

the feet for ornaments and had lopped them off. Any wounds were concealed beneath the puffy tufts of feathers. The natives then handed the footless carcasses to European naturalists, whom they never imagined could be so naïve. "The love of the marvelous, and the fondness for conjecture prevailed," said one historian, and a scientific myth was born.

* Although they likely had some scientific training, ship surgeons didn't always know the ins and outs of collecting specimens. So one London collector helped out by providing starter kits with jars for insects and special paper for pressing plants. He also included unconventional advice in letters to his aides, including the importance of rooting through the digestive tracts of predators for half-digested species: "Whenever you catch any of these," he emphasized, "look into their gutts & stomach & take out Animalls you shall find there." Actually, this is good advice even today: in 2018, scientists in Mexico discovered a new species of snake inside the bowels of another.

and tutoring. He failed at all of it, and he seemed destined for a dead-end career. He finally got a lifeline in the summer of 1771 when a physician and botanist named John Fothergill announced a specimen-collecting expedition to Sierra Leone. Fothergill was a Quaker and a determined foe of slavery. He nevertheless compromised and sent Smeathman to a slave colony because there were no other settlements in Sierra Leone to choose from.

Despite his own qualms about slavery, Smeathman jumped at the offer, since science was a well-trod path to becoming a gentleman then. Part of the incentive was social. If he played his cards right, he might be elected a fellow of the prestigious Royal Society. There were financial incentives, too. Each of Smeathman's three main sponsors put up £100 ($12,000 today) to finance the trip. In exchange, they got to select £100 worth of the specimens he would send back. After that, Smeathman could sell the rest for profit. Arrangements like this were not uncommon for aspiring scientists from lowly families. Eighty years later, Alfred Russel Wallace, co-discoverer of the theory of evolution by natural selection, would engage in a similar hustle in Malaysia.*

In January 1772, a few weeks after arriving in Africa, Smeathman set up a home base on the Banana Islands, a cluster of two and a half sandy spits off the coast of Sierra Leone. (At high tide there were three isles; but the low tide uncovered an isthmus between two of them, for an average of two and a half.) He spent a few weeks on the Bananas recovering from the first of several bouts of malaria, then went to see the headman of the islands, the colorful James Cleveland.†

---

*Some historians have suggested that Wallace's work in Malaysia might even have pushed him toward his co-discovery of evolution by natural selection. After all, being a collector involved scrutinizing thousands of bugs for variations in color, size, and other traits, and variation is the raw material on which natural selection works.

†James Cleveland's father William had come from a respectable family in England; William's brother was Secretary of the Admiralty. But William had a

With Cleveland's blessing, Smeathman built an English-style home on one island, complete with a garden. Cleveland also secured the Englishman a wife. The young woman—thirteen years old, Smeathman estimated—was the daughter of a local chief. Mixed-race marriages like this were common in Africa, but unlike many Europeans, Smeathman mooned over his wife. "The nuptials were celebrated by above one hundred discharges of cannon from the shore and . . . the only bull within many miles was killed on the occasion," he bragged to one sponsor. "My little Brunetta with her wooly toppin is laid in bed beside me . . . Gad so! I believe I am in love with her! She [has] . . . a shape like the Venus of Medicis, with two pretty, jutting, dancing hills upon her breast." This admission of affection is startling. Most Europeans wanted sex and grub from their women, and nothing more.

Marrying the daughter of a chief also brought Smeathman under the chief's patronage and protection. This in turn allowed him to recruit local African freemen as guides and start his scientific expeditions. For the most part these expeditions consisted of tramping about the countryside and grabbing plants and animals to ship back to England. There, they'd be dissected and classified according to Linnaeus's taxonomic system, the dominant paradigm of the day. But Smeathman also went beyond such work and did pioneering research in ecology and animal behavior. This included his studies of the legendary termite mounds of western Africa.

---

rascally streak, and after being shipwrecked near the Bananas in the 1750s, he staggered ashore and declared himself king. He married several local women there and eventually sired James, who built a robust slave-trading business despite being half-African himself. To keep Cleveland happy, Europeans on his island had to supply him with a steady stream of guns, rum, cloth, and iron goods, not to mention the odd golden belt buckle or decorated drinking horn. At one point, Henry Smeathman ordered for Cleveland a wildly expensive "electrical machine" from England, which built up charges (presumably with friction) and used a glass orb to shock people for amusement.

The intricate interior of a termite bugga bug mound, as pointed out by one of Henry Smeathman's local freeman guides. Notice several men standing on a mound in the background. (Drawing by Henry Smeathman.)

These mounds—known locally as bugga bug hills—stood like small volcanos on the African plains, steep cones up to twelve feet high. Although made of little more than dirt and termite spit, they were sturdy enough for five grown men to stand atop one, and were considered the best vantage point for watching ships enter local harbors.

To study the mounds, Smeathman and his guides would sneak up with hoes and pickaxes and rain down blows on the mud walls. Then they'd claw at the broken dirt with their fingers and scramble for a glimpse of the interior. This haste was necessary because, within a few seconds of the first blow, they'd hear an ominous crackling sound—"shriller and quicker than the ticking of a watch," Smeathman recalled. It was an alarm call, and a moment later several brigades of termites would erupt out of the hole and attack. The bugs had a vicious bite, and sent the barefoot guides howling for cover. The Europeans fared better at first, but inevitably the termites would wriggle inside their shoes and chomp down, staining their white stockings with bloody red polka dots. (A true man of science, Smeathman later used the stains as data, estimating that an average termite loosed a quantity of blood equal to its own weight with each bite.)

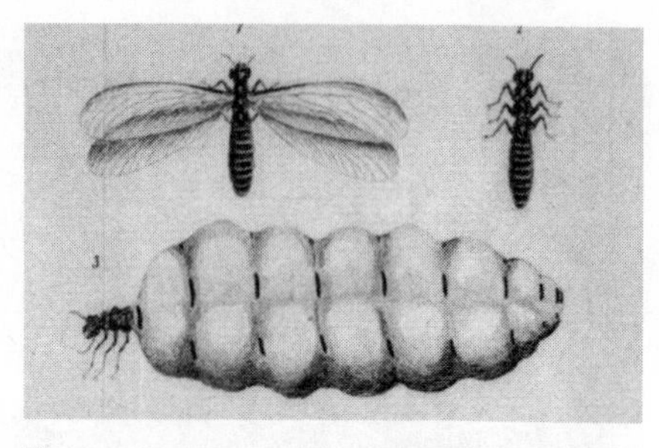

The grotesque termite queen (*number 3*), a tiny torso grafted onto an egg-laying sac that pumps out 80,000 ova every day. (Drawing by Henry Smeathman.)

Smeathman would soon become a weary expert on the pain of different insect bites, but his forbearance allowed him to study the interior of bugga bug mounds in incredible detail. In fact, Smeathman's account of them reads like an architecture primer, with references to turrets, cupolas, naves, catacombs, flying buttresses, and Gothic arches. He also speculated (correctly) that the shape of the mounds acts as a bellows, pumping in fresh air and keeping the insides at a consistent temperature. Enthusiastically, if a bit chauvinistically, he declared that each termite mound "gives a specimen of industry and enterprise as much beyond the pride and ambition of men as St. Paul's Cathedral exceeds an Indian hut."

The termites themselves also fascinated him. It no doubt cost him many a nip on the hand, but Smeathman finally dug deep enough into the mounds to reach the "royal apartment" and glimpse the grotesque termite queen.* She was little more than a tiny torso grafted

* Analogies can help us understand scientific systems sometimes, but using the name "queens" with regard to termites, ants, and bees is misleading. These queens aren't the "rulers" of the colony in any meaningful sense. In fact, a queen's life seems pretty miserable. When establishing a new colony, workers essentially wall the queen into a tiny "royal chamber" where she lives out the rest of her life in the dark, unable to do anything more than force food down her gullet and pump out babies all day long. Imagine being in perpetual labor for the rest of your life, and so bloated that you can't walk or even drag yourself around. Rather than think of them as queens, a more apt name might be the royal gonad.

onto a three-inch-long, pulsating egg sac, which pumped out eighty thousand ova every day, nearly one per second. (He estimated that the queen weighed thirty thousand times her average subject, the equivalent of a five-million-pound pregnant woman.) Other termites were no less amazing. In one chamber, Smeathman discovered small white pellets that he mistook for more eggs. But under his microscope, they revealed themselves to be tiny mushrooms. To his shock, he realized the termites were raising crops for food. Scientists now know of several other animals that do the same, but Smeathman was the first person to realize that *Homo sapiens* were far from the first farmers in Earth's history. (Ants, in fact, have been farming for sixty million years.)

In doing this research, Smeathman was following in the footsteps of the German naturalist Maria Merian, once known as the "mother of entomology" for her pioneering studies in Suriname in the late 1600s. (Being wealthy, Merian paid her own passage to South America, showing a remarkable independence of spirit. While she did employ slaves to collect for her, she at least acknowledged their help in her papers, unlike most naturalists.) Merian was among the first scientists to study the full life cycle of insects, including the foods they ate at each stage. She was also a gifted artist, and recorded some gruesome scenes in her notebooks, including one where a tarantula the size of a sasquatch paw pins down a hummingbird and feasts on it.

In the same spirt, Smeathman studied termites from eggs to adulthood, and made several drawings of termite mounds that are still celebrated today for their dramatic flourishes. The pictures are intriguing sociologically as well. Rather than portray himself as the hero at the center of the action, Smeathman shows his African guides smashing open the bugga bug hills, tacitly crediting them for their help. Historians have also noted that, unlike later reproductions of the drawings, Smeathman's originals don't alter the guides' features to conform with European standards of beauty; his men are recognizably African.

All this stands in accord with the general respect that Smeathman paid his guides. Rather than scoff at their knowledge of natural history, as most Europeans did, he let them correct him on some points—like the fact that a certain winged termite he'd seen wasn't a separate species, but a stage in the life cycle of an already known species. (Linnaeus himself got this wrong.) Even more remarkable, considering how deep this prejudice runs even today, Smeathman put aside any disgust he might have felt and indulged in the local practice of eating insects. His guides showed him how to skim termites off pools of water and roast them like nuts over a fire. As Smeathman wrote, "I have eat[en] them dressed this way several times, and think them both delicate, nourishing, and wholesome. They are something sweeter, but not so fat and cloying as the caterpillar or maggot."

To be sure, Smeathman shared some of the biases of his day. At different points in his letters, he describes Africans as overly "cunning" and "full of sloth and full of villainy," among other insults. But he's much harsher on European slavers, calling them "beasts" and "monsters" and the "outcasts of France, Holland, Denmark, [and] Sweden." He also respected local Africans' knowledge of medicine; they had "valuable secrets in the vegetable way." He even praised their oratory skills after observing them at their local courts, dubbing them "black Ciceroes and Demosthenes" who in many ways surpassed English barristers.

Smeathman's work on bugga bug mounds eventually won him respect among European biologists—as well as a delightful nickname, Monsieur Termite. And if that's all there was to Smeathman's story, he would have gone down in history as both a sharp scientist and a tolerant and forward-thinking fellow. Unfortunately, there is more to unpack here. Smeathman's guides were mostly local freemen, not slaves. So during his first months in Sierra Leone, he could insulate himself from slavery and console himself that his ties to it were minimal, involving just trade and transport. But keeping his distance proved harder than expected. As his initial funds ran low, he began to make friendly with slavers in order to secure better trading terms. Then, gradually, he began to let his guard down around them for an all-too-human reason. He got lonely. By April 1773, his seventeenth month in Africa, he was openly lamenting his isolation in letters to his sponsors. Despite having three wives by then, his heart ached for the company of a fellow countryman, someone who spoke the same language and worshipped the same god and was stirred by the same hymns. So bit by bit, Smeathman began taking advantage of the slave-traders' offers of hospitality. He told himself this was just a palliative—temporary relief for solitude.

Natural history wasn't the only scientific field to exploit the slave trade. The first major astronomical observatory in the southern hemisphere, in Cape Town, was built with slave labor. Edmond Halley of comet fame solicited data about the moon and stars from slavers in different colonies, and geologists collected rocks and minerals in such places. The Royal Society sent questionnaires to slave ports asking for observations and profited from its investments in slave companies.

Even a field as rarified as celestial mechanics benefitted. For the most part, Isaac Newton was a solitary, stay-at-home crank—scribbling down equations at his desk and hiding them from colleagues. But when putting together *Principia Mathematica,* which includes his famous law of gravity, Newton made a radical and very public prediction: that the gravitational tug of the moon causes tides. To prove this, he needed data about the height and timing of tides from across the globe, and one crucial set of readings came from French slave ports in Martinique. Celestial mechanics is literally otherworldly, about as far removed from grubby human life as you can get. But slavery was such a fundamental part of European science then that not even the *Principia* could escape its shadow.

Still, there's no question that natural history benefitted the most from slavery—and in some cases even helped slavery expand its reach. Colonial merchants eagerly sought out natural resources abroad, like dyes and spices, and they consulted with scientists about the best way to hunt for and cultivate such goods. In addition, medical research into quinine and other drugs helped white Europeans survive in tropical locales. And the safer and more profitable a colony was for Europeans, the more that commercial activity there, including slavery, thrived. Scientific research, then, not only depended on colonial slavery but opened up new markets for it.

Some European naturalists in the Americas also forced slaves to collect specimens for them, especially in dangerous places. They sent slaves shinnying up trees or diving into frigid pools; other slaves negotiated tangles of thorns or fatally slippery slopes. Surprisingly, a few collectors actually paid slaves for their help—a half-crown for every dozen insects ($18 today) and twelve pence ($7) for every dozen plants, provided they weren't ragged. Most collectors were stingier, and the vast majority of Africans who gathered specimens received neither money nor acknowledgment. Only glimpses of these men and women survive today in plants like Majoe bitters, named after a gray-haired slave in Jamaica who used its bark to treat yaws, a syphilis-like skin disease.

The best-known African naturalist was Kwasi, a *lockoman* (sorcerer) who was active in Suriname in the 1700s. Although a slave himself, Kwasi often sided with white Europeans at the expense of Africans, and he remains a controversial figure even today. As one European observer noted, Kwasi crafted amulets out of "pebbles, sea-shells, cut hair, fish-bones, feathers, &c., the whole sewed up with a string of cotton round the neck." Kwasi then sold these amulets to slaves fighting for their freedom, assuring them that the magic inside would render them invincible in battle. It didn't, but that didn't stop Kwasi from profiteering. According to oral histories, he also infiltrated a troop of escaped slaves in the jungle, then betrayed their position to white soldiers. For deeds like this, Kwasi

A slave named David is forced to climb a tree and skin a boa constrictor for his master, John Stedman. Oddly enough, this drawing was by the poet William Blake.

was granted his freedom, as well as given expensive European garments, including a gold breastplate inscribed, "Quassie, faithful to the whites." In retaliation, one troop of escaped slaves ambushed him and hacked off his right ear.

However controversial, Kwasi was nevertheless regarded as a botanical genius. He was especially famous for a preparation of root powder that quieted stomach pain and quelled fevers. Many white Europeans actually submitted to medical treatment from him rather than trust their own doctors,* a striking vote of confidence. For thirty years, Kwasi refused to identify the root, until he finally led a disciple of Linnaeus into the forest one day and pointed to a shrub with vibrant red flowers. The disciple brought the shrub back to Linnaeus, who dubbed it *Quassia amara*. It's a rare example of a species named after a slave.

It's probably not a coincidence that Kwasi, so faithful to the whites, was immortalized by European scientists, while so many other talented men and women were lost to history. But for every European name attached to some plant or bug, it's worth remembering that there were probably one or two or a dozen unnamed helping hands.

Unlike Kwasi, Smeathman was no botanist. He was a bug guy, and trying to sort through all the unfamiliar flora in Sierra Leone left him frustrated and overwhelmed. He was thrilled, then, when a letter in early 1773 informed him that another disciple of Linnaeus, the botanist Andreas Berlin, would be joining him on the Banana Islands. Not only could he offload the botany, but he'd have another gentleman-scientist to keep him company.

Although just twenty-seven years old, Berlin already had an impressive resume, having sailed with Captain James Cook on one of his celebrated scientific voyages. It didn't take Berlin long to prove his worth as a botanist, either. On his first expedition with Smeathman, in April 1773, Berlin discovered three species new to European science within fifteen minutes, a haul that delighted him. "I am like

*In contrast to Kwasi, some slaves used their superior botanical knowledge to revenge themselves on their captors by poisoning them. Cassava was an especially popular poison because it's a delectable dish if cooked properly but toxic otherwise. Slaves would take worms that fed on cassava juice, dry them out and mash them up, and conceal the resulting powder beneath a fingernail. Then they surreptitiously dropped some into a bowl while serving their masters a meal.

a blind person who, having just had his eyes opened," he gushed in a letter, "sees the sun for the first time. He falls down in wonder . . ." For all his talent, however, Berlin had one major vice: liquor. Every hour not spent botanizing was spent boozing, which made Smeathman furious—especially because his other assistant was also a sot. "To have two assistants, and neither of them sober," he complained, "is rather unfortunate."

Smeathman's native helpers were giving him fits as well. Most were local villagers who snickered behind his back at his habit of picking up contemptible little bugs and weeds. Snickered, that is, until Smeathman announced that he was willing to pay for these specimens. After that, he had more "help" than he could handle: "Men, women, and children crowd in to stare, to ask questions, and bring things to sell: every plant with a flower . . . every commonest insect, even cockroaches and spiders in the houses." Smeathman eventually started turning people away, to their confusion and disgust. Some exacted revenge by stealing specimens from under his nose and making him pay for them twice.

Increasingly frustrated, especially with Berlin, Smeathman began blowing off steam through one of the few outlets he had—socializing with slavers.

To be sure, Smeathman never warmed up to the lowlifes who manned the slave ships—the crude, foul-mouthed sort who, as he once sniffed, stirred their tea with a "rusty, dirty, greasy knife" and ate butter so rancid that it was fit only to grease wagon wheels with. Instead, Smeathman cottoned onto the merchants and ship captains, the aristocrats of Sierra Leone slavery.

In truth, these "gentlemen" were every bit as cruel as the sea dogs. Worse, they were the ones actually profiting on slavery. But they had some polish to them, and Smeathman began dropping by their "country estate" on Bunce Island for games of whist and backgammon. He also played golf on the rugged, two-hole course there. (Smeathman called the game "goff," and it was a tad different than today. The ball was the size of a tennis ball, and the holes, he said, were "the size of a man's hat crown.") In a callous choice of words, Smeathman described golf as "a very pretty exercise for a warm climate, as there is nothing violent in it except the single blow" of the swing. Meanwhile, actual violence was taking place a quarter-mile away, on the far side of the island, where slaves were chained up in pens and flogged. In May 1773, Smeathman also went hunting for goats on the Îsles de Los, and bent his rules on imbibing to enjoy a grog-soaked feast on the beach. One of his companions on this lark was none other than John Tittle, the slave captain who would soon toss a boy into the sea to retrieve his hat and send a barrel of feces to the father. But for that day at least, he and Smeathman were chums.

Shortly after the feast, Smeathman hitched a ride back to the Banana Islands on Tittle's slave ship, and sketched a harrowing portrait of a disease outbreak there. One historian aptly described it as "Dantesque": "[T]wo or three slaves thrown overboard every day dying of fever, flux, measles, worms," Smeathman wrote. "Here the Doctor dressing sores, wounds, and ulcers, or cramming the men with medicines, and another standing over them with a cat[-o'-nine-tails, i.e., a whip] to make them swallow."

The victims of the outbreak included Andreas Berlin. Drink had already ruined his constitution, but even when laid low with fever and diarrhea, he demanded his daily ration of grog onboard the ship. (He also ate loads of pineapple, possibly as a folk cure.) Smeathman withheld the booze at first but soon gave in—to his regret. Berlin died shortly afterward, just three months into his African adventure.

After that blow, Smeathman leaned even harder on slavers for company. This descent into moral turpitude wasn't simple or linear; as the disease passage above shows, the man who'd been rendered mute with grief by the sight of two Black mothers nursing their infants was still there, and still recognized the evils of slavery. But the overall trend was unmistakably downward. At first he'd relied on slavers for material support only—equipment, food, mail. Then he made friendly with them to secure better trading terms. In time, making friendly led to actual friendships, to fight off the loneliness darkening his days. As any psychologist could have predicted, increased contact with slavers also led to sympathizing with their views, even defending them.

Things only deteriorated from there. A year and a half into Smeathman's expedition, very few specimens (beyond some insects) had reached England. This wasn't all Smeathman's fault. Specimens took time to prepare, and because the triangular trade ran in one direction only, his boxes had to be packed onto slave ships and returned to England via the Caribbean, delaying their arrival by months. Plus, ocean voyages weren't exactly the safest environment. If the sunlight, heat, humidity, or saltwater swells didn't destroy his specimens, then the worms, ants, and rodents onboard usually did.

As a result, Smeathman's empty-handed sponsors began grumbling about their poor investment in him. He in turn realized that his reputation as a scientist—and his hopes of becoming a gentleman-scientist—would be in shambles unless more specimens started reaching them, and soon. To this end, Smeathman began working as an agent for a Liverpool-based slaver, to help grow the slaver's business in Sierra Leone. In exchange, Smeathman secured space for his specimens on the rare ships returning straight from Africa to England. Preserving dead bugs and plants meant more to him than preserving his morals.

By mid-1773, Smeathman was dabbling in the slave trade himself. Hard currency was somewhat useless in Africa, as people there preferred to barter with goods—including slaves. A captain delivering some packages to Smeathman from England, for example, once demanded a slave as payment. The local economy ran on slaves as well. As Smeathman rationalized it in his letters, he was chronically short on "candles, sugar, tea, butter," shoes, nails, and other necessities. However much he regretted this fact, slaves were a sort of universal currency in Sierra Leone, the one "commodity" that he could trade for anything. This included goods like tobacco and rum that he needed to pay off chiefs and secure guides. Without their help, he would have had to suspend his scientific expeditions—an idea he would not countenance. So he began trading slaves for goods when necessary.

Predictably, by 1774, Smeathman had moved beyond merely trading slaves locally to selling them to plantations in the Americas, to help fund his research. Smeathman would continue to defend his participation in the slave trade in letters: the economic reality of life there, he insisted, pushed him into the market. But his conscience does break through here and there. In one passage, he confesses, "My scruples in regard to the slave trade are vanished." He'd become part of the system he despised.

It would be nice to believe that the scientific sins of Henry Smeathman's day are dead and buried—after all, the transatlantic slave ended in the early 1800s. But the truth is, we owe our modern, scientific worldview to books like *Principia Mathematica* and *Systema Naturae,* both of which depended on slavery for their completion. Even more acute, many specimens collected via the slave trade still reside in museums to this day.

The most important museum goods trace back to Hans Sloane, a London doctor and naturalist.* As a young man, Sloane collected on plantations in Jamaica, and he later married into a rich slave-owning family. Using that wealth, Sloane then bought up collections from other naturalists, and he eventually amassed the largest natural history collection in the world, tens of thousands of items. Disturbingly, these included human specimens, as he recorded in a private catalogue: "the Skin of the arm of a black injected [with] red wax & mercury"; "the foetus of a negro from Virginia"; "stones extracted

*Curiously, Hans Sloane also invented milk chocolate in Jamaica; he considered it an easier way to consume cocoa, which was considered medicine then. Upon returning to London, Sloane sold the recipe to an apothecary, who in turn sold it to a little outfit called Cadbury. Every time you nibble a chocolate bar today, you can trace it right back to the scientific-slavery-industrial complex.

from the vagina of a negro African girle." Sloane used this collection as a springboard to become president of the Royal Society in 1727, taking over for none other than Isaac Newton.

Upon his death in 1753, Sloane did something unusual. He wanted to provide financially for his daughters, but he also wanted to keep his collection intact, rather than let it get dispersed at auction. So in his will he offered everything to the British government for £20,000 ($3.1 million today) to establish a museum. To raise the money, the government set up a lottery with £3 tickets ($470), and despite some shady dealings—including the organizers buying tickets in bulk and scalping them—the lottery raised £300,000 ($47 million). Because government officials wanted the museum to serve the public, they called it the British Museum. It quickly became one of the most renowned institutions on Earth. Later, most of Sloane's items were transferred to the Natural History Museum in London, another beacon of civilization. Sloane's specimens—many of which had a direct link to slavery—thereby became founding collections for some of the most famous cultural institutions in the world.

To be fair, there's no reason to single those museums out. Other specimens linked to the slave trade can be found in Oxford, Glasgow, and Chelsea. In fact, almost every natural history museum in any major European city—Paris, Madrid, Vienna, Amsterdam—probably has items of similar origin. These aren't just dusty curiosities, either. Scientists still consult these collections to study plant domestication and historical climate change. They also extract DNA from specimens to study how plants and animals have evolved across the centuries. Most scientists, however, remain oblivious to the origins of the items they use.

Even many historians remain ignorant. But some, at least, unable to look away any longer, have started to untangle the origins of museum collections. A few even want to bring science into the larger discussion about slave reparations and the cultural legacy of slavery. As one put it, discussions about the profits of slavery are usually framed "in terms of just dollars and cents, pounds and pence. Yet [the profits] can also clearly be measured in specimens collected and papers published."

Acknowledging this legacy can be painful for scientists. After all, isn't science progressive—a force for good in the world? Absolutely. But it's also a human endeavor, full of well-meaning but fallible people, people who get fixated on their research and ignore their nagging consciences. People like Henry Smeathman.

Ultimately, Smeathman's compromises got him what he wanted in science—up to a point. Dabbling in slaves secured him enough supplies and trade goods to make several long expeditions to the bugga bug mounds, and he collected so many specimens overall that one sponsor later complained about the glut: "My house could not possibly contain one half." By late 1775, after four years in Africa, Monsieur Termite felt sure enough in his scientific reputation to return to England, where he imagined that a hero's welcome would await him. So he packed up his specimens and booked passage for the Caribbean aboard the slave ship *Elizabeth*.

The moment Smeathman stepped onboard, the captain seized his personal chest of bugs and plants and dumped everything out. The captain then packed the ships' pistols inside, since the chest had a good solid lock to keep the guns secure in case of mutiny or a slave revolt. But the captain soon had bigger things to worry over, since the *Elizabeth* leaked like an old roof and required constant pumping to stay afloat. (A few weeks after their arrival in the West Indies, she was condemned as unseaworthy.) Of the 293 slaves aboard the *Elizabeth,* 54 died en route to the Americas.

Smeathman had planned to depart for England immediately after arriving, but another bout of malaria had wiped him out on the journey over, and he didn't want to face the harsh winter winds on the return Atlantic voyage. He decided to rest for a few months instead. By the time he felt shipshape, however, the Revolutionary

War had broken out, and American privateers were seizing British ships left and right. Suddenly marooned, Smeathman ended up settling in Tobago and spent the next four years doing natural history on different islands. Most notably he studied Caribbean fire ants, which were sweeping through various islands in swarms so large that Moses himself would have hesitated to call them down on Pharaoh. The ants attacked even animals on the islands, skeletonizing horses and cows overnight. Locals referred to the swarms as ant "blasts."

Mostly, though, Smeathman spent those years ruminating over slavery. The West Indies were idyllic in many ways—lush, green, full of novel specimens—and he spent many a happy day tramping about collecting flora and fauna. Every so often, though, he'd wander near a plantation, and the crack of a whip would rend the air, followed by a scream. He also witnessed public whippings of slaves, both men and women, and the long, twisting scars that crisscrossed their bodies haunted his sleep. (Slave owners often dripped candle wax or rubbed chili peppers into the wounds to make the sting worse. Some put chilis directly into slaves' eyes.) In Africa, Smeathman could still keep slavery at a distance. But the cruelty of plantation life righted his moral compass and he disavowed slavery once again.

Smeathman finally traced the last leg of the triangular trade and sailed for England in August 1779. Naturally, pirates seized his ship on the way and dumped all his remaining specimens—years' worth of labor—into the sea. He returned to England broke, and the hero's welcome that he'd envisioned never materialized. He did present a well-regarded paper on termite mounds to the Royal Society. But the snooty president of the society decided that Smeathman wasn't enough of a gentleman for their ranks and effectively blocked his election as a fellow. Smeathman was no doubt heartbroken, his dream of becoming a gentleman-scientist dashed.

Instead he struck out on his own and became a scientific lecturer, speaking to sold-out crowds about his adventures with ants and termites. Smeathman became a bit player in the abolition movement as well. In fact he always ended his scientific lectures with a short sermon on slavery, "that infamous policy," as he once put it, "which degraded one species [i.e., race] of human beings to pamper the luxury of a few of the others."

Perhaps feeling guilty about his slave-trading days, he also began raising funds to start an agricultural colony in Sierra Leone for free Blacks. This included loyalist slaves in North America who'd fought with the British against their masters during the Revolutionary War. Hundreds of men and women signed up, including dozens of mixed-race couples who simply wanted to live somewhere free of harassment. Smeathman even traveled to Paris to meet Benjamin Franklin and seek the famous American's endorsement for the plan. (While there, Smeathman happened to observe the world's first balloon flight in 1783, courtesy of the Montgolfier brothers. The spectacle inspired Smeathman to design his own, cigar-shaped balloon with wings, which he hoped would prove more steerable than the spherical Montgolfier one.)

In July 1786, however, just months before the settlers planned to depart for Africa, Smeathman was felled by another bout of malaria. South American countries were still hoarding quinine then, and a mere three days later—before anyone could procure some for him—he died. Four hundred colonists still set sail later that year, but they arrived in the middle of the rainy season, and without Smeathman's contacts and expertise, they had to beg food to survive. Within three months, a third of them were dead. Eventually a local chief evicted the remaining colonists and burned down their shacks, sending Henry Smeathman's grand dream of redemption up in smoke.

Despite his premature death, Smeathman did advance the abolitionist cause in one real, albeit indirect, way. In early 1786 he'd written a tract about his vision for the Sierra Leone colony, and two Swedish scientists who read it, a mining engineer named Carl Wadström and a botanist named Anders Sparrman, were inspired enough to travel to Africa themselves in late 1787. They had vague plans to visit the interior of the continent, but ended up stranded at a French slave port in Senegal. What they saw over the next few months appalled them—and unlike Smeathman, they didn't stick around long enough for their outrage to erode.

Instead they stormed back to London and began regaling people with tales of "slave dungeons" and men and women "lying chained in their own blood." They also revealed a diabolical French scheme to capture slaves on the cheap. Rather than risk their own necks in raids, the French would sell arms to two rival tribes and provoke a war between them. Inevitably, one side would take their enemies prisoner, at which point the French would sweep in and buy up the captives. Wadström described the aftermath of one such war, when the victorious tribe marched into port with the soon-to-be slaves, singing and clapping and blowing horns: "Taking in the shrieks and agony of the one, and the shouts and joy of the other, with the concomitant instruments of noise, I was never before witness to such an infernal scene." Perhaps most scandalous of all, Sparrman and Wadström hadn't had to dig much to expose these machinations—the French slavers practically bragged about them, proud of their cleverness.

The Swedes ended up appearing before the House of Commons and British Board of Trade, and their testimony caused a sensation in London—both for what they revealed and also because of who they were. This was the 1780s, the high tide of the Enlightenment, and scientists at that time were considered beyond reproach—utterly unimpeachable witnesses on the big issues facing society. (Different times . . .) As a result, many people who'd hesitated to condemn slavery before were suddenly swayed to the abolitionist cause. Because if *scientists* said the slave trade was evil, who were they to argue?

To be sure, the Swedes didn't end the slave trade in the British Empire by themselves. Africans did a lot. Freed slaves like Olaudah

Equiano and the Sons of Africa provided their own damning testimony, and the long, bloody, and ultimately successful slave revolt in Haiti in the 1790s left the British public questioning what their government was supporting. The Quaker church also deserves credit for its long, lonely fight for abolition. But as leading abolitionist Thomas Clarkson said, as soon as the Swedish scientists went public, "The tide . . . which had run so strongly against us, began now to turn a little in our favour." In this way, Wadström and Sparrman helped science redeem itself a little after its long entanglement with slavery, and become a positive force for ending it.

Smeathman died before he could fulfill his dream of becoming a member of the august Royal Society. It must have been especially galling to know that, in skipping over him, the society had tapped other scientists with highly dubious reputations. One in particular, a medical doctor and rough contemporary of Smeathman's, led the most flagrant organized-crime spree in science history, robbing hundreds of graves to procure bodies for anatomical dissections.

In fact, doctors deserve their own special section in the annals of sinful science. Because physicians work directly with human beings, they often give science a human touch. But working with people also introduces new ethical dilemmas and new opportunities for abuse.

3

GRAVE-ROBBING: JEKYLL & HYDE, HUNTER & KNOX

The murders started so innocently. In a snug stone boardinghouse in Edinburgh, nestled beneath the shadow of the city's famous hilltop castle, an old man named Donald was clinging to life. He had fluid on his lungs (dropsy) and was essentially drowning on dry land. When he finally passed away one night in November 1827, his landlord William Hare arranged for a church burial.

But then Hare got to thinking. The church couldn't pick up the body right away, and Hare mentioned to a neighbor, William Burke, that he wished he'd kept his mouth shut and sold the body instead. At that time, possessing and selling a dead body was not illegal, and there was a strong, if seedy, market for them: anatomists in Edinburgh always needed corpses for dissections, and they paid ready money. Burke agreed that this was a missed opportunity. But rather than sulk, the duo simply made their own luck. A carpenter soon came by to seal Donald in a coffin, and left the two men alone with it. Acting fast, they wrenched the lid open with a chisel, hid Donald's corpse in a nearby bed, and filled the coffin back up with refuse—dead weight. When church officials stopped by later to claim the coffin, they were none the wiser.

Now the pair had to unload the body. They wandered over to a medical school, but the chief anatomist was absent. So they went to one of his rivals, Robert Knox. Knox was also absent, but his assistants told the duo to come back later; so they bundled Donald up that night and schlepped him over for Knox to evaluate. The famed anatomist was bald on the crown of his head, and his left eye was blind from a smallpox attack. He was also something of a dandy, although if he'd been working that night, he would have been dressed in a blood-smeared smock.

Burke and Hare laid Donald out on a green-felt dissecting table in Knox's lab and unwrapped him. Then they held their breath while Knox cast his good eye over the corpse. The tension must have been excruciating—would he suspect they'd stolen it?

I'll give you seven pounds, ten shillings, Knox finally said.

The pair took the money and scrammed. Burke felt guilty, but no one had gotten hurt. They needed the money anyway.

But like money always does, that £7s10 evaporated pretty quickly. And when an old miller named Joseph showed up at Hare's boardinghouse a few months later, and fell deathly ill with fever, the pair couldn't help but start thinking again. Hare was anxious to remove Joseph no matter what: He didn't want his place getting a reputation as pestilent. And given that the old man was practically dead anyway, why not nudge him along? No one knows who first proposed it, or whether they even dared suggest it aloud. But before another day had passed, Burke found himself gently pressing a pillow to Joseph's face. Hare then lay across the miller's chest to still his lungs. Just like that, they had a new body to sell.

Serial murderers William Hare (*left*) and William Burke (*right*). (Drawings by George Andrew Lutenor.)

Or did they? Going to Knox this time must have been doubly nerve-racking. Surely an expert anatomist could spot a homicide.

Burke and Hare needn't have worried. As every murder-mystery fan knows, strangling someone's neck will usually break the hyoid bone there, since that bone is fragile and cracks under pressure. But the Edinburgh duo's method of smothering the face and chest—soon to be known as "burking"—left the hyoid bone intact. In other words, they'd stumbled into a devilishly clever way to suffocate someone.

Given the state of forensic science at the time, it would have taken a determined eye to find evidence of murder—and if anything, Knox was determined *not* to find such evidence. Like all anatomists in that era, he knew not to ask questions about where his specimens came from. But in accepting Burke and Hare's bodies, he was helping to kick off the most deadly crime spree in science history.

There's a widespread myth that the Christian church in Europe banned human dissection long ago, driving the science of anatomy underground. In reality, churches in Italy often worked with anatomists, storing bodies for them after last rites. Church officials even encouraged the dissections of would-be saints. How else would they get at the bones and hearts and other shriveled relics that brought in pilgrims and packed the pews? Other countries were equally tolerant. One French playwright complained that staged, public dissections drew such big crowds that they were cutting into the audiences for his shows. By the 1600s scientific dissections were fairly common throughout Europe.

At least continental Europe—Great Britain did ban dissections. People there feared that dissection after death would leave their bodies mangled on Judgment Day, when God would resurrect the dead. Prudish Brits also viewed dissection as shameful—a naked body lying there, poked and prodded. Secular officials, however, were the ones who enacted the bans, not priests and bishops.

Still, the British government did supply some bodies to anatomists. These were usually executed criminals, who were sentenced to "death and dissection" as a "further terror and a peculiar mark of infamy." But even in an age when cutting down the wrong tree could get you hanged (really), there were never enough executions to satisfy the demand from medical schools. (Today, two medical students usually split one body in their preliminary anatomy class; back then, if they'd relied solely on legally donated bodies, the ratio would have been several hundred to one.) This shortage in turn led to unseemly scenes at public hangings, with students from rival medical schools brawling over corpses. In their haste, they sometimes even yanked people off the gibbet who weren't quite dead yet. Their necks hadn't broken and they'd merely passed out from lack of air—only to pop awake later on the dissection table. Others weren't so lucky. One latter-day review of dissection records found that the heart was still beating in ten of thirty-six cases. But at that point, it was too late to turn back.

During the dissections, the students would slice the bodies open at the abdomen with a knife and pick through the individual organs and tissues inside. They'd study where the main blood vessels ran, what the liver was connected to, how nerves threaded into muscles, and so on. This gave them a better idea of how the body worked and how its parts fit together, the very foundation of a medical education. Otherwise, you had doctors trying to identify diseased organs without knowing what healthy ones looked like, an impossible task. Worse, without a detailed knowledge of anatomy, doctors were liable to sever an artery or nerve while digging around inside someone, leaving the patient paralyzed if not dead.

Given the shortage of bodies for dissection, British anatomists (and their counterparts in North America) felt they had no choice but to rob graves. Some scientists did the deed themselves, while others enlisted students to help, swearing them to silence at the beginning of the semester like some necrophiliac fraternity. The oaths rarely worked, though. "Under cover of the night," as one observer put it, "in the most wanton sallies of excess," the students would liquor up and storm the churchyards to unearth fresh bodies. To them, it was all a macabre game.

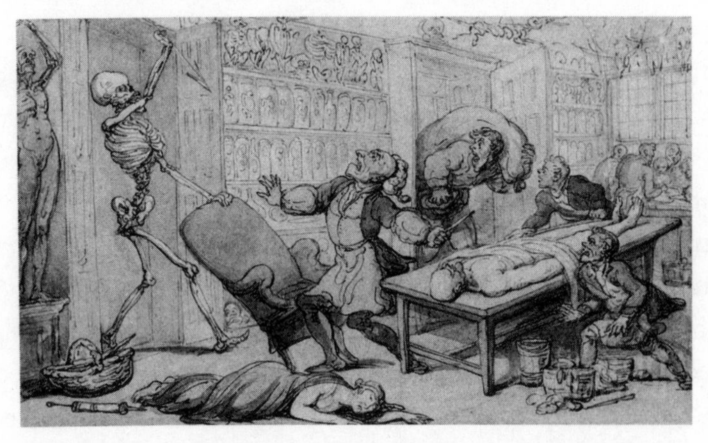

A parody of a dissecting room. Notice the resurrectionist entering with a bundled-up body. (Painting by Thomas Rowlandson.)

Government officials tended to look the other way at grave-robbing, for two reasons. First, most government officials were rich and powerful. Most bodies for dissection, meanwhile, came from the pauper class. Officials could therefore tolerate grave-robbing without the fear of their own loved ones going missing. Less cynically, authorities also knew that budding doctors and surgeons needed bodies to train on—and frankly, make mistakes on. Otherwise, the tyros would be learning anatomy on the fly inside live patients, making mistakes while elbow-deep inside their guts. Many government officials wanted to legalize dissection for this reason, but popular opinion prevented it. As a result, the British medical community fell into an uneasy truce when it came to procuring bodies. Don't ask, don't tell.

What finally broke the equilibrium was one man's obsession. John Hunter was the William Dampier of anatomy, revered for his discoveries and reviled for his methods. Coarse and foul-mouthed, with hair so red you could light a cigar with it, Hunter was the tenth of ten children in his Scottish family, and he went into medicine in part because six of his siblings died young from disease. He also had a role model in his brother William, an obstetrician in London who was highly praised (and highly paid) for discreetly delivering the children of Important Men's mistresses. William also taught anatomy on the side, but didn't want to sully himself carving up bodies. So in 1748, at age twenty, Hunter moved to London to become his brother's assistant dissector. He'd never cut into a body before that, but after the rush of that first incision, he basically never stopped.

Hunter's obsession took two forms. First, he loved anatomy for its own sake, and not just human anatomy. He carved up thousands of animals, too, including outré bits like "sparrows' testicles, bees' ovaria, and monkey placenta"; he even collaborated with Henry Smeathman in dissecting his grotesque termite queens. Second, Hunter saw anatomy as a way to reform medicine. Medicine in that era paid lip service to things like observation and experiment, but everyday treatments still consisted of hoary nostrums like purging, bloodletting, and tobacco enemas—literally blowing smoke up someone's ass. Hunter wanted to modernize medicine, and saw anatomy as the foundation of reform: to cure disease, doctors needed intimate knowledge of the body. To him, this included not just how the parts fit together but the feel and smell and even taste of different tissues. He once described the gastric juices of cadavers as "saltish or brackish." More daringly, he reported that "semen . . . when held some time in the mouth . . . produces a warmth similar to spices." Hunter even dissected and tasted an Egyptian mummy.

Whether in spite of or because of his unorthodox methods, Hunter made dozens of anatomical discoveries, including the tear ducts and olfactory nerve. He oversaw the first artificial insemination in humans and pioneered the use of electricity (from crude batteries) to jump-start the heart. He also charted the development of babies in utero and divined the modern classification of teeth into incisors, cuspids, bicuspids, and molars. Based on such work,* Hunter was elected to the Royal Society in 1767. Moreover, his practiced cutting hand and intimate knowledge of anatomy made him a celebrated surgeon. He eventually bought a house in London with a grand façade for receiving distinguished patients such as Adam Smith, David Hume, William Pitt, and Joseph Haydn.

Still, Hunter had his critics—especially for his dealings with grave-robbers. Most anatomists despised "resurrectionists" and "sack-'em-up men" as lowbrow thugs. In contrast, Hunter's vulgar manners actually made him a great favorite with grave-robbers. His majestic house even had a second, less wholesome, back entrance just for resurrectionists; it overlooked an alley, and at 2 a.m. they'd slink up and unload that night's catch. As one student remembered, the rooms back there were distinctly "perfumed" with the scent of corpses. Robert Louis Stevenson used this Janus-faced house, and Hunter's life in general, as models for *Dr. Jekyll and Mr. Hyde.*

Anatomist, surgeon, and grave-robbing abettor John Hunter served as a model for the novel *Dr. Jekyll and Mr. Hyde.* (Painting by John Jackson.)

* In perhaps his most clever discovery, Hunter settled a longstanding dispute about digestion. Many scientists back then argued that the stomach digested food either by applying heat to break it down or by mechanically churning it. But after noticing holes in his cadavers' stomachs, Hunter argued for chemical digestion. After death, he reasoned, the body stops producing the protective mucus that lines the stomach, and the acid there begins to digest the organ itself. This explained the holes, which neither the heat nor mechanical theories could. We now know that mechanical churning does play a role in digestion, but chemical action is primary.

Grave-robbers usually worked in teams. Less sophisticated crews would poach from mass graves, the open pits that were left unattended until they'd filled with paupers' bodies. The best crews had more elaborate setups. Many of them employed female spies—who attracted less attention—to linger near hospitals and workhouses waiting for people to die. The spies would then attend "the black" (thief cant for a funeral) and follow the wake to the "hospital crib" (graveyard) to note the location of the plot. Spies also kept an eye out for booby traps, such as spring-loaded rifles buried in the dirt and "torpedo" coffins that exploded if tampered with. Less drastically, some families would arrange twigs, stones, or oyster shells into a pattern on the surface of the plot, so they could tell if the dirt had been disturbed. The lady spies passed all this information on to the gangs for a cut of the proceeds.

The actual resurrecting took place at night. Sack-'em-up men had to become amateur astronomers, in fact, and chart the rising and phases of the moon to determine the time of peak darkness. Guards were little worry. If a graveyard even had one, the gangs either bribed him or got him so drunk he passed out. Then the thieves would tiptoe up to the fresh grave, disable any booby traps, memorize the pattern of sticks or shells, and start digging with their soft, quiet, wooden shovels.

The gangs rarely disinterred a whole coffin—too much work. Rather, they'd expose just the head of it, then jigger a crowbar underneath the lid and use the weight of the overlying dirt to snap the boards. A rope slipped under the arms of the body retrieved the prize. Brutally, they often disfigured the face at this point to prevent recognition. Before leaving, they stripped the shroud and any jewelry off the body and discarded it, since stealing gold or clothing would ratchet the deed up to a capital offense. Pros could empty a grave in fifteen minutes, and they were veritable Picassos when it came to recreating the look of an undisturbed plot. More than one gang snuck into a churchyard and started digging, only to find an empty grave below—the work of a more punctual rival.*

(Resurrectionists had other tricks for making money, too. Rather than sully themselves with digging, some went the confidence-man route. They'd visit an almshouse or hospital, pick out a body there, and start weeping and rending their garments as they claimed their dearly beloved "uncle" or "great aunt." As a variation, some gangs would sell a body to an anatomist and then enlist a confederate to knock on his door an hour later, before dissection began. Posing as a relative, the confederate would demand the body back, on threat of calling the police—at which point the whole gang would walk down the street to another anatomist and sell the body again. Even more brazen, one gang wrapped their very-much-alive friend in a sack and sold him to an anatomist. They were apparently hoping the anatomist would set the sack aside overnight—at which point the friend would hop out, rob the house blind, and sneak off. The plan was thwarted when the anatomist realized the "corpse" was still alive.)

Gangs earned flat fees for adult bodies, around £2 in Hunter's day—roughly what farm laborers made in a whole season. For "smalls" (children), gangs charged by the inch.† For rare specimens

*People who died during the summer were lucky in that their bodies decayed faster in the heat. Hence, they were of less use to anatomists, who often took summers off. Wintertime deaths almost always meant a body-snatch. On especially cold days, when the bodies were still stiff, the resurrectionists didn't even have to conceal them. They could simply prop them up in carriages like passengers and drive them right up to the anatomist's back door. In other cases bodies were concealed in shrouds or sacks, or even shipped around in barrels labeled "pork" or "beef." This might explain the line about "beef" in the nursery rhyme later in this chapter.

†Anatomists dissected infants and children whenever possible in part because they were eager to chart the course of human growth and development, a hot scientific topic at the time. More practically, infants made for handy teaching specimens. To study the nerves and blood vessels, anatomists had to pump colored

(e.g., pregnant women during their last months), prices might rise to £20 ($2,500 today). One industrious grave-robber once cleared £100 in a single night.

However lucrative, the work had its dangers. If caught, resurrectionists risked jail time or transport to the colonies.* And while the police often looked the other way, mobs didn't: Grave-robbers regularly got beaten, shot, or whipped with metal wires. One horde, displaying a keen sense of irony, tried burying a grave-robber alive in the pit he'd just dug. Some anatomists acted like godfathers and looked after their most reliable resurrectionists, bailing them out of jail or providing for their families during stints in prison. But if anatomists double-crossed them, or bought bodies from a rival crew, the gangs had no compunctions about breaking into labs and hacking the bodies up, rendering them useless for dissections. It was straight Mafia tactics. *Pretty little corpse you got there. Be a shame if something happened to it.*

Hunter, however, rarely ran afoul of resurrectionists, mostly because he couldn't afford to: all his research depended on them. Later

---

wax or mercury throughout the body, sometimes by blowing it with a pipe. And it was a heck of a lot easier to pump fluids throughout a tiny child's body than a full-grown adult's.

Incidentally, anatomists dissected the bodies in a strict order, based on how quickly different tissues went putrid. The lower abdomen was first, since those organs turned foul fast. Then came the lungs (which were often black from London's sooty air) and the heart. Muscles decayed more slowly, so they could be put off. Finally came the bones, which anatomists would sometimes wire together to form skeletons. Despite the rush to get through the putrid parts, dissecting rooms often reeked of rotten flesh—as did the anatomists. In order to keep them focused on their studies, medical schools often banned surgical apprentices from marrying, but given where they spent much of each day, you wonder whether the ban was necessary.

*Just to clear up some of the legal subtleties: Possessing a dead body wasn't a crime; no one could technically own a body, and corpses weren't considered property. That said, resurrectionists could still get nailed for violating a grave, which was illegal. And again, stealing clothing or jewelry on the body was definitely a crime, often punishable by death.

in life he estimated that, during the dozen years he worked for his brother, he dissected or observed the dissection of two thousand corpses—one body every two days.

Given that nearly every one of those bodies had been stolen—sometimes by Hunter himself—this was bad enough. But month by month, corpse by corpse, Hunter also developed a moral callus, and pretty soon these former human beings became nothing but bags of bones to him. Probably the most disgraceful episode involved the Irish giant Charles Byrne.

Byrne was so tall—eight-foot-four, according to the tabloids—that people swore he could light his pipe from gas streetlamps without rising to his tiptoes. Scholars at the time ascribed his fantastic height to his parents having sex atop a haystack; modern doctors suggest a pituitary tumor that pumped out excess growth hormone. To earn a living, Byrne exhibited himself in county fairs across Ireland and England, wearing gigantic frilled cuffs and a three-cornered hat the size of a topsail. He had an audience with King George once, and the moment John Hunter laid eyes on Byrne, he grew obsessed with dissecting him.

To this end, Hunter approached Byrne one day in London and offered to buy his corpse pre-posthumously. To Hunter, the offer was an honor. Who wouldn't want to be dissected by the world's leading anatomist? (No hypocrite, Hunter later had his own assistants carve him up after his death.) But Hunter's obsession had blinded him to the fact that most people considered dissection an abomination, and Byrne practically shrieked at the offer. After sending Hunter away, the giant gathered his friends and made them swear to God above that they'd dump his body in the sea when he died, to keep it out of the anatomist's clutches.

Sadly for Byrne, death came sooner than expected. Pituitary conditions can cause arthritis and bad headaches, and he reportedly started drinking to blot out the pain. (Hunter learned this through a spy he'd employed to tail the giant from pub to pub.) It would have taken prodigious amounts of booze to get Byrne sozzled, and his liver eventually sputtered out. He finally drank himself to death in June 1783, just twenty-two years old.

As one newspaper reported, anatomists began circling Byrne's house "just as Greenland harpooners would an enormous whale." Byrne's friends ordered a coffin the size of a schooner and, figuring that Byrne had exhibited himself when alive, put him on display in death and began selling tickets. True to their word, however, no one got the body. After four days of cashing in, they and an undertaker began a seventy-five-mile march to the sea to fulfill the dearly departed's last wishes.

Unfortunately, the mourners had more in the way of good intentions than good sense. Lugging a giant coffin around was hard, sweaty work in the June heat, so the Irish lads began stopping every few miles to refresh themselves with ale and toast their friend. Being responsible fellows, they always tried to bring the coffin inside the tavern with them to watch over it; when it didn't fit inside, they made arrangements to keep it safe. At one tavern, for instance, the door was too narrow for Byrne's bier, so they took the suggestion of the undertaker and stored it in a nearby barn he knew of. Eventually this nomadic wake reached the coast past Canterbury, where they engaged a local bark and rowed out to the deep. There they pushed the coffin of the Irish giant off the prow, and watched it sink to the bottom of the sea.

The Irish giant's *body,* meanwhile, was back in London. Before the wake had set off, Hunter's spy had approached the undertaker with a £50 bribe for his cooperation. The undertaker, sensing desperation, soon drove that offer up to an incredible £500 ($50,000 today). Hunter couldn't afford that, but his mania got the better of him and he agreed. The undertaker then steered Byrne's friends to the tavern above with the narrow door, knowing the coffin would never fit. He'd already bribed the owner of the nearby barn to let him hide some tools and men among the straw inside, and while Byrne's friends made merry, the undertaker's crew unscrewed the lid, hid the giant in the straw, and replaced him with a precisely measured weight of paving stones. Afterward, the coffin went one direction, the body another. By dawn the next morning, Hunter was dragging the giant through the Mr. Hyde entrance of his home.

Strangely enough, he never dissected Byrne. Had he, his trained eye might have spotted the pituitary tumor and connected it to gigantism, a link that remained undiscovered for another century.* Hunter, though, was scared enough of Byrne's friends to abandon his plans. Instead, he focused on boiling the body down to preserve the skeleton. He used a huge copper vat to do so, skimming off the fat like so much soup and picking out the giant's bones. Hunter eventually opened a museum of anatomical oddities in London (one writer called it "Hunter's collection of human miseries"), where the seven-foot-seven-inch skeleton served as the centerpiece. Against the giant's wishes, it's still on display today.

Hunter left behind two conflicting legacies. There's no question he was one of the great scientists of his day, making dozens of new discoveries about how our bodies work. And beyond any specific findings, he inaugurated a new spirit in medicine, dragging it out of the realm of bloodletting and tobacco enemas and emphasizing observation and experiment, a big step toward scientific respectability. He also inspired countless students (Edward Jenner and James Parkinson, to name two), and enrollment in medical schools boomed after his death in 1793.

That said, Hunter's lack of ethics gravely undermined his reputation. Condemning scientists in the past for not living up to today's moral standards is unfair, but even in his own day, people despised Hunter. In a neat trick, he managed to make enemies of both patrician doctors, who recoiled at his rubbing elbows with body-snatchers, and the plebian masses, who resented being fodder

*The famed neurosurgeon Harvey Cushing finally opened up the Irish Giant's skull in 1909 and found clear evidence of a tumor. Namely, he noticed that a structure called the sella turcica—a saddle-shaped notch in the base of the skull that houses the pituitary gland—was enlarged in Byrne, a common occurrence in giants.

for his research. Even Hunter's fellow anatomists blanched when he stole Charles Byrne's body. He's a classic example of someone rationalizing his sins by pointing out all the good that resulted, as if ethics was merely moral accounting, with the good deeds canceling out the bad.

Worse was to come. More than anyone else, Hunter transformed grave-robbing from the wanton sallies of students into an industry, and the sheer number of bodies he bought distorted the market for them. The boom in medical-school enrollment further exacerbated the shortage of bodies and drove the price still higher, from roughly £2 in the 1780s to £16 (nearly $1,000) in some places by the 1810s—equal to what an average laborer made in five years. To be sure, Hunter was no monster. However flexible his conscience was, he at least had one. But the higher the price for bodies rose, the more that people without any scruples at all were tempted to jump into the game. People like Burke and Hare.

The memory of smothering the old man with a pillow tormented William Burke. He took to gulping whiskey at night just to fall asleep, and kept a bottle on his nightstand for reinforcement. William Hare was less troubled. The old man would have died anyway, so why sweat it?

Given their circumstances, however, neither man gave back the money they'd earned. Burke, then in his mid-thirties, had grown up poor in Ireland and fathered a child when young. He'd eventually moved to Scotland alone to support his family and taken various dead-end jobs there—digging canals, soldiering, baking bread. His wife back home finally stopped answering his letters, and he'd moved in with another woman in Edinburgh. Hare's background was sketchier. He was probably younger than Burke and had likely also immigrated from Ireland. While Burke had a round, warm face, Hare had narrow eyes and that lean, hungry look that Shakespeare warned against. For a few years, Hare had been helping to manage his wife Margaret's boardinghouse, but they were barely scraping by. Burke, who worked as a cobbler, was struggling, too. So, troubled conscience or not, when Burke ran low on money again, Hare had little trouble convincing his friend to give murder another go.

A fairly inaccurate dramatization of a Burke-Hare murder. Their victims were almost always passed out drunk, and both men helped kill them—not by strangulation, but by sitting on the chest and plugging the mouth and nostrils, a murder method now known as "burking." (Engraving by Robert Seymour.)

An elderly woman named Abigail Simpson took a room at the boarding house in mid-February 1828. The duo got her so drunk she vomited, but they kept plying her with porter and whiskey until she passed out. Honestly, she might have died of alcohol poisoning at that point, but Hare lay on her chest to be sure, and Burke pinched her mouth and nostrils shut until she fell still. Simpson's body likely fetched around £10, and although Burke reverted to gulping booze at night, it was a little easier to sleep this time.

It would soon get a lot easier. As Burke once put it, the pair figured "we might as well be hanged for a sheep as a lamb," and over the next ten months they went on one of the biggest murder sprees in history, burking fourteen more victims. They killed an old woman and her mentally handicapped grandson. They killed another old woman with a single tooth in her mouth, then a daughter who'd reportedly stopped by to search for her. With two victims, they never even learned their names. At first, the duo simply waited for good candidates to visit the boardinghouse, but they eventually got itchy and started luring people in. Burke, the chatty one with the warm face, would linger near liquor stores in the early morning, scouting for down-on-their-luck alcoholics who needed a daily eye-opener. Then he'd win their confidence and invite them over to Hare's for a warm meal and more drink. When the dupes finally passed out, the Williams sprang. Burke remembered that the victims "would convulse and make a rumbling noise in their bellies" as they sank toward death. All the bodies then made their way to anatomist Robert Knox.

Although not as brilliant as John Hunter, Knox was a talented scientist, and he was vastly more couth. During lectures he wore handsome coats and shirts trimmed with lace, and his fingers, while stained red, were studded with diamond rings. Still, he shared Hunter's appetite for human flesh, and he faced stiff competition for bodies in Edinburgh, where hundreds of new medical students arrived each year. Given that pressure, it was all too easy to accept bodies from anyone who knocked. In the words of a later nursery rhyme about the trio, "Burke's the butcher, Hare's the thief / And Knox the boy who buys the beef."

The notorious Dr. Robert Knox, who "bought the beef" from the murderers Burke and Hare. (Courtesy of Wellcome Trust.)

To be sure, Knox's assistants had their suspicions about Burke and Hare. One actually confronted Burke once with some hard questions about where a certain corpse came from. (Burke shot back, "If I am to be catechized by you, where and how I get subjects, I will inform the Doctor [Knox] of it!" The assistant backed down.) Even if the assistant had informed Knox, Knox might not have done anything. Any competent anatomist could have seen signs of asphyxiation in the Burke and Hare bodies: bloodshot eyes, flushed faces, a telltale trickle of blood near the mouth. But the intact hyoid bone gave Knox plausible deniability. Most of the victims reeked of booze anyway, and it was sadly common for alcoholics to suffocate on vomit. In short, Knox closed his lone eye to any sign of trouble, hesitant to upset such reliable suppliers and interrupt his research.

The more "beef" Knox bought, the more reckless Burke and Hare became. One day Burke saw two policemen harassing a drunk woman and chivalrously offered to escort her home; he steered her to the Hares' instead and burked her. The most ballsy murder involved Daft Jamie, a beloved "town idiot," who wandered the streets barefoot and was known by sight to everyone. The duo snuffed him out anyway and trundled him over to Knox. And instead of burning Jamie's clothes, as they'd done with other victims, they gave the clothes away to friends; several articles were recognized around town by their former owners, who were puzzled. When Knox and his team gathered to dissect Daft Jamie, an assistant took one look at the face and gasped. A tight-lipped Knox said nothing, and ordered them to prep the body.

These close calls only emboldened Burke and Hare, and their murder spree culminated with a triple homicide plot around Halloween in 1828. This time the guests—a young couple named Ann and James Gray and a petite, forty-something Irishwoman named Margaret Docherty—weren't staying with the Hares but with Burke and his common-law wife. (Burke had reeled Docherty in at a grocery shop by claiming that his name was also Docherty.) In an effort to dispatch Docherty first, Burke made several rather transparent excuses to send Ann and James away. Hare then met Burke at the latter's place. As usual, Burke and Hare got Docherty drunk; perhaps nostalgic for home, they also induced her to sing some Irish ditties. Things then swerved unexpectedly. Sometime around 11 p.m., Burke and Hare got into a violent quarrel and Burke began strangling his junior partner. Docherty screamed, "Murder! Murder!" and an upstairs neighbor alerted the police.

This being Halloween, however, a night full of mischief, the police were busy elsewhere. No one stopped by. And when Burke and Hare finally disentangled themselves, they turned their murderous rage toward Docherty and burked her. Afterward, they stripped off her red gown and hid her body in some straw at the foot of a bed.

Incredibly, Burke let Ann and James Gray back into his house the next morning, probably with an eye toward murdering them, too. But his behavior made Ann—the hero of this whole story—suspicious. Burke clumsily spilled whiskey several times, as if to mask an odor, and when Ann offered to tidy up his house, Burke refused. She noticed in particular that he never let her near the straw at the foot of one bed.

Finally, late on November 1, All Saints' Day, Ann found herself alone in the house and made straight for the straw. She suspected Burke and Hare had pulled a Halloween heist of some sort and were hiding illicit goods there. Instead she found an arm, which was attached to a naked woman with a trickle of blood on her lips. Ann grabbed her husband and fled, but they met Burke's common-law wife Helen at the door. Helen offered them money to stay mum, but Ann and James pushed past her and ran for the police.*

The police quickly realized, however, that this wasn't an open-and-shut case. Yes, there was a dead body, but Burke and Hare could always claim Docherty had gotten drunk and choked. So in a bit of gamesmanship, the police weighed the two men's characters, decided Hare had fewer scruples, and offered him a plea bargain. It worked like a charm. Hare turned King's evidence, and in return for testifying against Burke, he escaped all charges.

Burke's trial started in late December and ran for twenty-four hours straight, ending inevitably in a guilty verdict. The judge

*Despite her heroics, Ann Gray's story does not have a happy ending. Her husband James died within a few months of their run-in with Burke and Hare, and as happened to so many widows then, she was basically left indigent.

sentenced him to hang. Hare, meanwhile, walked out of the courtroom a free man—albeit in disguise, since a mob was waiting to exact revenge. He fled like his animal namesake, and after some close calls in different towns, he finally escaped Scotland and disappeared, his last years every bit as mysterious as his early ones.

Burke was hanged on a rainy morning a month later. The death itself was unremarkable, although every last window in the buildings surrounding the jail was packed with faces. In a satisfying twist, his body was then handed over to Robert Knox's biggest rival for dissection and display in a museum. Macabrely, the rival even dipped his quill in the blood from Burke's skull to write out a placard: "This is written with the blood of Wm Burke, who was hanged at Edinburgh on 28th Jan. 1829 . . ."

Knox came within a spade of being indicted himself, but the evidence just wasn't there: he could still claim he didn't know. A mob in Edinburgh made an effigy of him anyway, "bald head and all," as one contemporary recalled. Rather than burn the effigy, the crowd burked it instead.

Outrage over the Burke and Hare murders (as well as some copycat homicides in London) finally forced British officials to do something about the lack of cadavers available for dissections. Specifically, they introduced a law to give anatomists unclaimed bodies from poorhouses and charity hospitals, bodies that no family members or friends stepped forward to claim. This would not only expand the number of bodies available for training and research, it would undercut the black market for them and allow scientists to cut ties with thieves, thugs, and grave-robbers.

But however tidy this solution seemed, using unclaimed bodies raised ethical issues of its own. In particular, poor people hated the plan, since they would still be the ones supplying most of the corpses. After all, it wasn't the well-heeled or well-connected who were dying unclaimed in almshouses.

In a callous response to this complaint, one politician argued that supplying bodies for research was the least the poor could do. After all, look at all the free meals and medical care they'd enjoyed on the public dime during their lives. (A rival politician countered that he too supported the dissection of those who were sucking the public teat dry. He proposed starting with the royal family.) More compassionately, some supporters of the law pointed out that, despite the unfair burden of supplying bodies, improving the training of doctors would benefit the poor more than any other group. For one thing, diseases often hit the poor much harder. The rich could also afford experienced doctors and surgeons, while the poor were stuck with greenhorns who would be fumbling around and making mistakes. Given that, better the greenhorns do so on dead paupers than living ones. Allowing the dissection of unclaimed bodies, in other words, was the lesser of two evils and would ease suffering for the poor overall.

In the end, such arguments won the day, and Parliament passed the Anatomy Act in 1832. But while the act eased tensions in Great Britain, it did nothing to quell resentment in the United States, where anatomists had always been loathed and "anatomy riots" were a regular feature of life. One anatomy department in particular—at that most renowned of American institutions, Harvard University—got dragged into a deliciously tawdry scandal when an illustrious alumnus went missing, and turned up somewhere he shouldn't have in expertly dissected pieces.

4

MURDER: THE PROFESSOR AND THE JANITOR

According to legend, America's first anatomy riot started with a crass joke. One afternoon in April 1788, a medical student at New York's General Hospital was dissecting a woman's body in a lab there. Suddenly, he realized he wasn't alone. A gang of street urchins had gathered at the window outside—gaping wide-eyed at a real-life dead person.

This annoyed the student, who wanted to work in peace. So to spook the boys, he reportedly grabbed the cadaver's arm and waved it at them. *Yoo-hoo!* Then he hollered, "This is your mother's arm. I just dug it up!"

Har har. Unfortunately, one of the boys had indeed just lost his mother, and he ran home to his father bawling. The father in turn grabbed a shovel and marched out to his late wife's grave. He found exactly what he expected inside—nothing—and he was furious.

He wasn't the only one. Body-snatching had always hit the poor harder than the rich. Rich folks could afford robbery deterrents like mortsafes, iron cages that surrounded coffins and made them hard to pilfer. The rich could also afford private guards to watch over their loved ones for the week or two it took a body to get too putrid to dissect. The poor had no such safeguards, and certain groups were hit especially hard in the United States: American Indians; Black people, both enslaved and free; and German and Irish immigrants. So when the boy's father returned from the graveyard and proposed storming New York General, he found plenty of pissed-off neighbors willing to join him.

When the mob arrived at the hospital, hundreds strong, the doctors and anatomists fled in panic; one hid up a chimney. The rioters proceeded to drag all the medical equipment out to the street to smash it. They also burned anatomical specimens and reburied several bodies in various states of decay.

Trashing the building didn't slake the mob's anger, however. Their numbers only swelled overnight, and the next day they marched on another medical building, at Columbia University. Alexander Hamilton himself had to stand on the steps and plead with them to stop. In the meantime, New York's mayor had jailed several medical scientists for their own safety. Undeterred, the mob—five thousand people by then—gathered before the prison. They proceeded to smash its windows and tear down a fence, howling, "Bring out yer doctors!" At dusk, the terrified mayor finally called in the militia. He also begged local political leaders to come by and help restore order.

However tense, things still might have ended peacefully if not for what happened next. Among the political leaders called in was John Jay, a future Supreme Court justice and future governor of New York. But his pleading did no good. What did a blue blood like him know about having his loved ones' graves robbed? Someone flung a rock at him, cracking his skull.

Another leader called in was Baron von Steuben, an army general and one of the heroes of the Revolutionary War. He too got beaned in the skull with a brick. As von Steuben staggered backward, bloody and dazed, he reportedly called on the mayor to have the militia fire.

Now, technically this wasn't an order. But the soldiers were already spooked and didn't need any more encouragement. When they heard a general yell "fire," they snapped up their rifles—and opened up on the crowd. Estimates vary, but by the time the smoke cleared, up to twenty dead bodies lay in the street. The riot had started over one corpse, and ended with many more.

New York was hardly an aberration. At least seventeen American anatomy riots took place before the Civil War, in Boston and New Haven, Baltimore and Philadelphia, Cleveland and St. Louis. And again, while the burden of grave-robbing fell mostly on the poor, rich folks weren't exempt. In Ohio, U.S. Senator John Scott Harrison—the son of former president William Henry Harrison and father of future president Benjamin Harrison—was dug up, stripped naked, and laid out for dissection before his family swooped in and saved him.*

Eventually, most American states passed anatomy acts (a.k.a. "bone bills") modeled on the 1832 bill in Great Britain. These laws gave medical schools the right to unclaimed bodies from hospitals and poorhouses. But the bills kicked up the same ethical issues in America as they did across the Atlantic. What's more, it soon became clear that using unclaimed bodies was not only ethically dicey but scientifically dubious, too. Because as crazy as it sounds, your income can affect your anatomy.

These differences trace back to hormones. There's of course lots of individual differences among poor people, but in a broad, general sense, poor people suffer from chronic stress at higher rates than those in the middle and upper classes. The reasons are obvious. Poverty-stricken populations generally have more medical problems, and fewer means to treat them. They're exposed to more pollutants,

---

*Even Abraham Lincoln was the target of a body-snatching plot, albeit not for anatomical reasons. On election night in 1876—a night chosen because most people would be distracted with the news—several felons broke into Abe's vault to snatch his bones and hold them for ransom. In addition to money, they also wanted to use the bones as leverage to help spring a good buddy of theirs from jail, an expert counterfeiter. Unfortunately for them, the Secret Service had infiltrated their gang with a spy, and the plot was foiled.

too, and especially back in the 1800s, many of them faced eviction and starvation on a regular basis. The body responds to such stressors by releasing adrenaline and other hormones, and chronic stress can affect the size and shape of the glands that pump these hormones out. Some glands, like heavily worked muscles, swell in size. Others exhaust themselves and shrivel. And because the poor alone were undergoing dissection then, the doctors learning anatomy on them had a skewed view of what those glands should look like. There was systematic error in their science.

This wasn't just an academic worry, either. It had real, deadly consequences.

In the 1800s scores of babies started dying from what we now call SIDS—sudden infant death syndrome. Naturally, doctors wanted to know the cause, so they started performing autopsies on SIDS babies. They noticed that most SIDS babies had one gland in particular that looked enormous, the thymus glands in their chests. In reality, these were normal thymus glands. They only *seemed* large compared to the wilted thymus glands that doctors usually found in babies from poor families. These poor babies had often died of chronic and stressful ailments like diarrhea or malnutrition. SIDS babies, in contrast, died suddenly by definition—before diarrhea or malnutrition could wither their glands down. As a result, their thymus glands were normal-sized.

Unaware of all this, pathologists began blaming SIDS on hypertrophied thymus glands, which they decided were crushing babies' windpipes and suffocating them. So to shrink the glands down, doctors in the early 1900s began blasting them with radiation. Thousands upon thousands of children suffered burns, depleted glands, and, later, cancer as a result, leading to an estimated ten thousand premature deaths. It's a poignant example of how an unethical scientific setup can lead to dangerous scientific outcomes.

Eventually, the voluntary donation of bodies eliminated the need to use unclaimed corpses. Philosopher Jeremy Bentham, the founder of utilitarianism, became the first person in history to donate his body to science in 1832, in part to lessen the stigma of dissection. His good deed didn't convince many at the time, but the world had come around to Bentham's thinking by the mid-1900s. Today, the majority of cadavers dissected in medical schools are gifts.

Still, medical schools today often struggle to find enough cadavers. One analysis from 2016 found that New York City medical schools fell three dozen bodies short of the eight hundred they needed to train new doctors, a 5 percent gap. In other states, the gap is closer to 40 percent. Countries such as India, Brazil, and Bangladesh face even bigger shortfalls. Nigeria has close to 200 million people, yet some medical schools there get zero annual donations. To make up for shortages, latter-day resurrectionists are digging up buried bodies again or swiping them from funeral pyres and selling them on the "red market."

It's not just whole bodies anymore, either. Like car thieves chopping up automobiles for parts, grave-robbers can make more money—up to $200,000—by hacking bodies up and selling individual tissues: teeth, eardrums, corneas, tendons, even bladders and skin. Often the families of the deceased have no idea this is happening; some have fetched their loved ones from funeral homes to find the bones replaced with PVC pipes. (At least they got the bodies back whole. In 2004, a funeral director from Staten Island got caught selling bodies to the U.S. Army for $30,000; the army was dressing the bodies in armored footwear and dangling them over land mines to test how well the footwear worked.) To be sure, the international laws governing transplant organs (lungs, livers, kidneys) are fairly robust and prevent such trafficking. But otherwise, as one anatomy professor lamented, "We are more careful with [importing] fruits and vegetables than with body parts." And while the poor are once again more at risk for being chopped up, it also happened to longtime *Masterpiece Theatre* host Alistair Cooke in 2004.

If all this makes you squeamish about the science of anatomy, you're not alone. Anatomists themselves continue to debate the ethics of different practices, and even in those cases where anatomical science does real good—bringing criminals to justice in murder cases, for instance, through forensic work—there's always a macabre undercurrent to the research. Much of forensic anatomy, in fact, traces its roots to a ghoulish case at Harvard Medical School in 1849. In many ways, it was a confrontation between the field's past and its future: the best minds in American medicine had to determine whether this was just another shady resurrectionist deal, or whether something more sinister had occurred.

It was the turkey that first got the janitor thinking *murder*. On Thanksgiving day in 1849, he had a succulent bird sitting on his kitchen table, a gift from his boss, Dr. Webster. Yet here he was, hacking at the brick wall of a latrine in the basement of Harvard Medical School. He wanted to be at home, feasting. But he just couldn't eat with all those clues nipping at his conscience.

The disappearing George Parkman. (Courtesy of the U.S. National Library of Medicine.)

The janitor was hardly alone in obsessing over the case; the people of Cambridge, Massachusetts, were talking about little else that November. Dr. George Parkman—tall and gangly, with a stiff, upright walk that left his chin jutting up at an impossible angle—had stopped by a grocery store one Friday afternoon to purchase some crushed sugar and a six-pound block of butter. He'd then asked the grocer to hold onto the goods, plus a treat for his invalid daughter: a head of lettuce, a delicacy in November. Parkman told the grocer he had an appointment to keep, and would be back in a minute to pick everything up. He never returned.

Parkman, near sixty, had graduated from Harvard Medical School in 1809, but he'd never practiced seriously. He'd preferred to amass real estate instead, and had actually donated the land on which Harvard's squat, three-story medical building sat. Less nobly, Parkman owned several tenement slums and was a stickler about rent. He also made a killing as a loan shark, hounding his debtors for every cent—especially if they'd crossed him.

And Dr. John White Webster had crossed him. Webster, fifty-six, was something of a hellion. He'd graduated from Harvard Med a few years after Parkman, and had done a residency in London, where he'd loved to attend public executions. "Hang at eight, breakfast at nine!", he'd cackle. He'd no doubt snatched a body or two in his day as well. But after practicing in the Azores for a while, Webster had given up medicine to teach geology and chemistry at Harvard; his lab sat in the medical building's basement. His lectures often featured pyrotechnics, and he loved whipping up laughing gas to get students high.

Harvard chemist and alleged murderer John White Webster. (Courtesy of the U.S. National Library of Medicine.)

If Webster had given up working as a doctor, though, he was still addicted to the doctor lifestyle. The typical Harvard professor then was independently wealthy and worth about $75,000 ($2.3 million today). Three-fourths were near 1-percenters, and some professors' mansions were so lavish that they appeared on local tourist maps, so people could walk by and gawk. In comparison, Webster's salary was $1,200, well below the university average of $1,950. Far from being a mere inconvenience, this penury actually put his job in danger. In the mid-1840s an Italian professor at Harvard had actually been forced to resign after going bankrupt; there were consequences for not living up to Harvard's social standards. So Webster chose to keep up the lifestyle he'd enjoyed as a doctor, buying a six-bedroom house in Cambridge with two parlors and entertaining lavishly with oysters and wine. But he couldn't afford servants—shamefully, his wife and daughters had to dust their own home—and his savings dipped so low that he once bounced a $9 check.

Rather than economize, Webster approached Parkman in 1842 for a $400 loan ($13,000 today). In 1847, he went back for $2,000 more ($62,000). Webster did try to make good on the loans over the next two years. But he had no financial discipline, and finally had to mortgage a beloved collection of minerals and gems to Parkman as collateral. People around town were soon whispering about Webster's debts, which infuriated him. While getting a haircut once, he heard an acquaintance joke, "Did you ever see a man shave a monkey?" It was probably an innocent crack, unrelated to Webster's finances. Webster hopped up anyway, snatched the barber's straightedge, and lunged. He just missed slashing the acquaintance.

By the fall of 1849, Parkman was pestering Webster for his money back, and the sheriff was threatening to repossess Webster's furniture. Desperate to buy time, Webster went behind Parkman's back and mortgaged his beloved mineral collection to two other creditors. Unfortunately, one of them was Parkman's brother-in-law, Robert Shaw. Shaw and Parkman happened to pass Webster in the street one day, and Shaw asked Parkman about Webster's finances. When Parkman asked why he was curious, Shaw mentioned the mortgage on the minerals. After a moment of confusion, they realized that Webster had essentially sold the same collection to both of them.

Parkman was livid at the news, and he eventually confronted Webster in the basement of the medical school. *Pay me or else,* he demanded. Both men lost their tempers, and the building's janitor overheard them quarreling—including Parkman's threat that "something must be accomplished." Webster finally promised to scrape together $483 ($15,000 today) and have it ready the Friday before Thanksgiving.

When that Friday came, Parkman paid for the butter and sugar from the grocer and dropped off the head of lettuce. His chin thrust up, he then marched over to collect from Webster. Webster later told the police that Parkman snatched up the $483 without a word and hurried off.

That's when the mystery started. Parkman was pretty OCD in his habits, so when he didn't turn up for dinner that night, his family began to fret. His absence the next morning panicked them. After some quiet inquiries, they put a notice in the newspaper offering $3,000 for information ($92,000 today). Seeing the notice, a chagrined Webster called on Parkman's brother and explained about their meeting. Hearing this, the family felt their stomachs crater. Parkman had a bad habit of carrying too much cash around after collecting on debts. He'd been mugged for it before and no doubt had been again, with fatal results. With a heavy heart, the family put a second notice in the paper offering $1,000—for Parkman's body.

In the meantime the police started dragging the nearby Charles River. They also roughed up some local hoodlums to wring information out of them. Nothing solid turned up. The last confirmed sighting of George Parkman was at the Harvard medical building. Indeed, there were rumors going around that Parkman's dog—who often accompanied him on debt-collecting rounds—had been seen lingering near the building, as if waiting for his master to emerge.

So a few days before Thanksgiving, the police made their way to the medical school to poke around. First they searched the apartment of the janitor in the basement, including underneath his bed. Nothing. Only then, and with great reluctance—they hated disturbing such an eminent scholar as himself—did the police go to the room next door and knock at Webster's office. A magnanimous Webster said he understood completely and let them in to search his lab. Or at least most of it. No one was brave enough to pick through his private latrine. They also found a locked closet, and when one cop asked what was inside, Webster explained that he kept explosive chemicals there. That ended that, and not long afterward the police bade the professor goodbye and returned to roughing up lowlifes. Little did they know, a much more obvious suspect had been right under their noses the whole time.

Harvard janitor Ephraim Littlefield. (Courtesy of the U.S. National Library of Medicine.)

Ephraim Littlefield was more than the medical school janitor. His chinstrap beard and high part made him look like a gentle Quaker, but he was also wrapped up in the dirty business of procuring bodies for anatomy classes. Because he and his wife lived in an apartment in the medical building's basement, he could meet resurrectionists at all hours. Littlefield wasn't above a little side action, either. A year earlier, a local doctor had botched a second-trimester abortion and killed his patient. Afterward, he tried selling both her and her dead fetus to a Harvard physician, Oliver Wendell Holmes Sr. In an unusual display of ethics for the time, Holmes said no. Desperate, the doctor turned to Littlefield and asked him to dispose of the corpses. Littlefield said he would—for $5. Littlefield got his money, but the doctor got caught anyway, and word of Littlefield's venality embarrassed the school.

This dark commerce should have made the janitor a prime suspect in Parkman's disappearance—and the police had indeed searched his apartment in the medical building. So perhaps to clear his name, Littlefield began his own investigation over the next few days, focusing on some nagging doubts he had about his boss, Dr. Webster.

Littlefield's basement apartment sat next to Webster's lab, and because the janitor's jobs included stoking a fire in the lab furnace each morning, he was used to entering and leaving the lab at will. Suddenly, after Parkman disappeared, Webster began locking the lab door. Yet the furnace inside was still blazing—so hot that Littlefield couldn't even touch the wall on the other side of it; he actually feared the room was on fire once. Even stranger was the turkey. Webster mostly ignored Littlefield as the help, and the professor was known to be in debt. Yet a few days before Thanksgiving he'd treated the janitor to an eight-pound bird. Why? And why had he made Littlefield trek across town to pick it up, instead of having it delivered? Was he getting Littlefield out of the way?

Suspicious, Littlefield began poking around. When Webster ignored his knocking on the lab door one day, the janitor dropped to the floor and, holding his breath, peeked beneath the doorway. He could just make out Webster's feet; the professor seemed to be dragging something toward the furnace. Later, Littlefield even slipped through an open window into Webster's lab, but a hasty search found nothing amiss.

That's when he decided to dig. On Thanksgiving Day, Littlefield found the medical building deserted. While his eight-pound turkey grew cold, he grabbed a hatchet and chisel, posted his wife as a lookout, and crept into the vault beneath the basement to hack through the brick wall of the privy there. The police had declined to search the professor's latrine. The janitor wasn't so squeamish.

He was a bit lazy, though. There were five layers of brick to the vault, and a hatchet just wasn't the right tool. So Littlefield quit after ninety minutes, cold and hungry. That night, perhaps to blow off steam, he went to a cotillion dance with his wife and stayed out until 4 a.m. He was pretty rusty the next morning, and had some odd jobs to do,* but he eventually dragged himself over to a nearby foundry to borrow a hammer, a better chisel, and a crowbar—to start work on a new water main, he claimed. Then it was down into the vault again.

He made swift progress for a while. Then he heard four hammer blows on the floor above him—*bang, bang, bang, bang*—his wife's signal that Webster was coming. Littlefield dropped everything and raced upstairs, only to find it was a false alarm. Still, Webster did show up not long afterward, and Littlefield had to play him off before he could return to the vault.

Several hours later he finally opened a hole in the innermost layer of brick. He held his lantern up, peering into the darkness. But a draft kicked up and almost snuffed the flame. (Given where he was digging, the draft no doubt slapped his face with some foul odors as well.) Still, Littlefield widened the hole and tried again, shielding the lantern this time as he reached it inside. It was a moment right out of Poe. He saw mostly what you'd expect in a latrine. But as his eyes adjusted to the gloom, he noticed one additional thing. In the center of the pit, glowing a dull white, sat a human pelvis.

Littlefield sprinted off and fetched the police, who finally did a thorough search of the lab. They found bone fragments and dentures

---

*One of Littlefield's odder jobs—the task that occupied him the morning after Thanksgiving—was moving around phrenology busts for an eminent professor at Harvard Med, John Warren. Readers of my book *Caesar's Last Breath* will recognize Warren as the surgeon who first championed anesthesia in the medical world. It's a great example of how scientists who seem so modern in some respects can seem so bizarrely antiquated in others.

That wasn't the only anesthesia connection, either. Both William Morton (the dentist-cum-confidence man who discovered anesthesia) and Charles T. Jackson (who claimed that Morton stole the idea from him) would later testify at Webster's murder trial. Jackson testified for the prosecution about some odd chemical splashes he'd seen in the Harvard medical building. Unbelievably, Webster and Jackson were allowed to chat with each other during a break in the trial, and Webster chided Jackson for opposing him. Jackson immediately offered to testify again—as a character witness for the defense.

among the ashes in the furnace, as well as several more shanks of leg in the latrine, which they hauled up on a plank. Most gruesome of all, one cop began digging through Webster's tea chest—where he kept his beloved minerals, the source of all this trouble. Near the bottom, the cop felt something squishy and decidedly un-rocklike. It was a gutted-out ribcage, with a left thigh jammed inside like a human turducken.

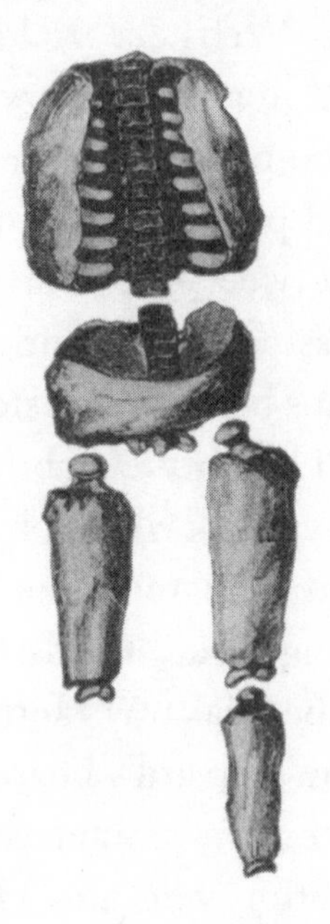

The remains of George Parkman recovered from the lab of John White Webster at Harvard. (Courtesy of the U.S. National Library of Medicine.)

The public was stunned. A murder—at *Harvard*? As one newspaper put it, "In the streets, in the marketplace, at every turn, men greet each other with pale, eager looks, and the inquiry, 'Can it be true?'" Poet Henry Wadsworth Longfellow, who taught Italian at Harvard and counted Webster as a friend, was crestfallen. "All minds," he lamented, "are soiled by this foul deed."

But if many were ready to hang Webster, local prosecutors looked at the evidence and swallowed hard. Just like with Burke and Hare, making a case wouldn't be easy. There was a body, sure, but it was headless. Was this actually Parkman? After all, corpses were shuttled in and out of the medical building all the time. Even if it was Parkman, perhaps someone had killed him elsewhere—or found him after a natural death and sold the body to the school. The police couldn't discount Littlefield, either. After all, who'd found the body? Everyone remembered his willingness to dispose of a dead mother and dead fetus for five lousy bucks, too. Perhaps he'd anticipated a reward for Parkman's body, or was in cahoots with some resurrectionists. Any way you sliced it, there was reasonable doubt everywhere.

Given the juiciness of the scandal, Webster's trial in March 1850 was likely the biggest court case in American history to that point. City officials actually built cattle chutes to herd spectators in and out of the courtroom; sixty thousand people tromped through over eleven days, and newspapers printed tweet-like hourly updates. The case also exposed fault lines of class and caste within the greater Boston area. Hardscrabble Bostonians blasted Webster as a psychopath and demanded that he swing. Haughty Cambridgians, meanwhile, sneered at Littlefield as a slimy sneak who'd obviously framed his boss. (Outside newspapers picked sides, too. One Virginia paper thundered over "that most disgusting of all bipeds, Ephraim Littlefield.") The presiding justice at the trial was Herman Melville's father-in-law. The judge also sat on Harvard's board of overseers, which normally would have been a conflict of interest—except that both the defendant and murder victim were also Harvard alums. As were the lead lawyers for both sides, plus twenty-five of the witnesses. It was half trial, half reunion.

Webster's defense was simple. *I'm a Harvard man, and Littlefield's not. So between the two of us, he obviously did it.* A bit more materially, Webster's lawyers pointed out that the state had no murder weapon and no idea how Parkman died. It was like a game of Clue: at different times the prosecuting attorneys suggested a sledgehammer, a knife, and Webster's "hands and feet" as the instruments of death. Could the jury really convict a man with no weapon and no visible wound on the corpse? Especially in a building full of corpses?

Still, the prosecution had one big thing going for it. The body had been found in a medical school, steps away from some of the world's foremost authorities on anatomy. They were experts at reading the human body, and while they respected Webster as a colleague, the corpse in his latrine told a damning tale.

First of all, the anatomists proved the body was Parkman's. Several of them had known Parkman for years, and their practiced eyes recognized the gaunt, trapezoidal torso discovered in the tea chest. Similarly, Parkman's dentist (another Harvard fellow) recognized the scorched false teeth from the furnace, since he'd crafted them himself. What's more, the dentist could tell that the teeth had been cooked inside a human head. If cooked in a furnace by themselves, he noted, false teeth would heat up quickly and pop like popcorn. These teeth hadn't popped, implying they'd been shielded from the heat by something moist, like human flesh. It was a virtuoso display of forensic dentistry.

As for who killed Parkman, the clues pointed to Webster. Whoever had carved up the body, witnesses noted, had betrayed an expert hand in separating the sternum, rib cage, and collarbone. Given the thick muscles and tendons in the breast, it was difficult to disengage the sternum without cracking it; only someone with practice dissecting bodies would know where to cut. The ex-doctor Webster fit that description, while Littlefield, for all his commerce in dead bodies, had never wielded a scalpel.

Still, despite all the evidence against Webster—the precise cuts; the scorched teeth; the fact that the body had turned up in his toilet—everyone knew that he'd get off, given the pro-Cambridge jury. The trial ended just before 8 p.m. on a Saturday, and the jury returned three hours later with the verdict. Melville's father-in-law hushed the courtroom and asked them how they found the defendant.

Webster had remained aloof throughout the proceedings, betraying no emotion. But when the word *guilty* rang out, he "started as if shot," said one witness, then slumped backward into his chair. A few yards behind him, Ephraim Littlefield broke down weeping.

Because of all the publicity surrounding it, Webster's trial provided a huge boost for forensic science in the United States, much like the O.J. Simpson trial familiarized laypeople with DNA evidence 150 years later. Equally important, after a century of riots and grave-robbing, the trial helped rehabilitate the reputation of anatomical science. Anatomists had not only collared a murderer, but in condemning the well-off professor and exonerating the poor janitor, they'd inverted the usual class alliance of anatomy. One observer, in fact, called the trial perhaps the fairest in American history: "There was never seen a more striking instance of equal and exact justice[:] money, influential friends, able counsel, prayers, petitions, the prestige of a scientific reputation, [all] failed to save him."

A dramatic recreation of the Webster-Parkman murder at Harvard.

And make no mistake, Webster did kill Parkman: he finally confessed to this a few days before he was scheduled to hang. During their final, fatal meeting, Webster said, Parkman had called him some dastardly names and threatened to get him fired—the last step toward financial ruin. In a fit of rage Webster had snatched a nearby log and smashed his tormenter's temple in. (As a former doctor, he apparently knew where to strike.) Parkman crumpled, and a panicked Webster dismembered the body and started burning it.

The confession, it turned out, was a last-ditch plea for clemency. As Webster told the governor's office, he'd committed manslaughter, not murder, and deserved jail time, not death. The governor was unmoved, and like William Burke before him, John White Webster was hanged for an anatomy murder a few days later.*

Despite the scandalous history of anatomy research, we can at least say this: The odd Burke-Hare murder aside, the people whose bodies were ransacked and dissected never felt a thing. It was still disgraceful, but at least they were beyond suffering.

Unfortunately, that hasn't always been the case. Most medical research takes place on the living, and as we'll see in later chapters, even the anatomists of the 1800s would have squirmed at some of the barbaric experiments that were coming in the following century. Human beings weren't the only ones to suffer, either. Medical research often treats animals as means, not ends, and their pain and anguish is brushed aside as collateral damage. That's a serious ethical quandary even when the experiments provide useful data. But in the case of Thomas Edison—who tortured horses and dogs with electricity simply to discredit a business rival—it strays into truly sinful territory.

---

*The notoriety of the case didn't end with Webster's death. The public couldn't get enough of the scandal and, bowing to necessity, Harvard eventually turned the crime scene into a tourist attraction. Littlefield became a local legend as well; souvenir hunters actually jumped him sometimes and snipped locks of his hair as mementoes.

The case had a long afterlife in people's memories, too. When Mark Twain visited the Azores in 1861, he was tickled to meet two of Webster's daughters, who'd no doubt moved there to escape their father's shadow. When Charles Dickens visited America in 1869, the one place he wanted to see in Massachusetts was the Parkman murder scene—to the mortification of locals, who assured him that the city had more to offer. Even as late as the early 1900s, a prominent Cambridge astronomer named Harlow Shapley could still get a big laugh with a joke about the case—that the most astonishing thing to him was that, in its whole long history, only *one* Harvard professor had murdered another.

5

ANIMAL CRUELTY: THE WAR OF THE CURRENTS

The crowd in the auditorium had no idea what they were about to witness, but the appearance of the dog put them instantly on guard. It was July 1888, at Columbia College in New York, and an electrician named Harold Brown dragged a seventy-six-pound Newfoundland mix onstage and forced it into a wooden cage surrounded by wire mesh. Sensing the audience's unease, Brown assured them that the dog was "a desperate cur, and had already bitten two men." A reporter on hand thought it actually looked meek—and was no doubt terrified.

While the dog cowered, Brown read a paper about the merits of alternating current (AC) versus direct current (DC), with an emphasis on how alternating current was deadlier. Upon finishing, he proceeded to do what everyone present feared, wrapping wet cotton around the dog's right forelimb and left hind limb, then wrapping the cotton with bare copper wire. The wire was connected to a generator, and when everything was ready, Brown flipped the switch.

Three hundred volts of direct current surged into the dog. It snapped into a rigid posture, and remained frozen until Brown killed the current. Brown then repeated the spectacle with higher voltages—400, 500, 700, 1000. After each pulse, the dog howled and quaked, and once slammed so hard against the cage that its head ripped through the wire mesh. "Spectators left the room, unable to endure the revolting exhibition," the reporter wrote. The dog's "vitality had been so reduced that it was a question with the audience whether he was dead or alive."

At this point, one spectator stood and demanded that Brown put the animal out of its misery. Brown archly answered that the dog "will have less trouble when we try the alternating current." Brown swapped the DC generator out for an AC one, and proceeded to zap the dog with 330 more volts—at which point, another reporter wrote, it "gave a series of pitiful moans, underwent a number of convulsions, and died."

One witness said the demonstration made a bullfight look like a petting zoo. Brown, meanwhile, was elated. He felt he'd proved his main point: that AC killed at lower voltages than DC. He knew this would be music to the ears of his benefactor, too, the man who'd sponsored the torture of the Newfoundland as well as several other animals—that American saint, Thomas Edison.

We all know the story. Despite less than three months of formal schooling, Thomas Alva Edison, through a mixture of gumption and genius, helped invent (or at least develop) dozens of innovative technologies—stock tickers, vote recorders, movie cameras, fire alarms, and more. His machine to record voices, the phonograph, so astounded people in the 1800s that many refused to believe it wasn't a magic trick. And while Edison didn't invent the lightbulb, he and his team of tinkerers did turn a dim, fragile, expensive fire hazard into a cheap, reliable device capable of illuminating the world. Edison absolutely deserves to be an American folk hero.

Thomas Edison and an early version of his phonograph—a brilliant invention that did not make him much money. (Courtesy of Gallica, Bibliothèque nationale de France.)

That said, Edison could be a real bastard sometimes. He and his assistants all put in grueling hours, regularly working past midnight and sleeping in closets at the lab. But Edison alone hogged the glory for "his" inventions. He was a backstabbing businessman, too. In the 1870s, Edison once accepted $5,000 ($110,000 today) from a telegraph company to develop some new electrical equipment. Edison did the job—then sold the rights to their competitor for $30,000. Even with the lightbulb, Edison lied several times in announcing publicly that he'd perfected it, both to spur investment in his company and destroy the stock price of natural-gas firms. Many people agreed with one executive who sneered that Edison "had a vacuum where his conscience ought to be."

For all his ruthless brilliance, though, Edison's inventions had one big flaw: they made very little money. Even the phonograph, however marvelous, was mostly used as a toy, since there was no market for recorded music then. Without steady income, Edison couldn't fund his true passion, his research lab. Moreover, Edison felt that, as a man of genius, he needed to transform the world somehow, and his scattered collection of gizmos wasn't getting the job done.

Finally, in the 1880s, Edison came up with his killer idea: wiring cities for electricity. Even at that time, the residents of most big cities walked around beneath a cat's cradle of wires strung overhead. These were mostly telegraph and arc-lighting wires, specialized for one purpose and restricted to certain businesses. Edison proposed threading electrical wires into every business, and even into people's homes. What's more, Edison's wires wouldn't be restricted to one purpose, but would supply power for everything—motors, looms, lightbulbs, you name it. Because Edison owned patents on every step in the process, from generators to transmission lines to consumer devices, all the profits from wiring cities would end up in his pockets. He also understood, like few contemporaries, just how revolutionary electricity would be—and he wanted to be the man to power America. He planned to start in Manhattan, perfect the technology, and expand to the rest of the country.

There was just one problem: His patents relied on direct current. Direct current is like a river, a flow of electrons in one direction only. Alternating current, in contrast, is like a fast tide: the electrons flow first one way, then another, alternating direction dozens of times per second. Both DC and AC can provide useful power, and for various reasons DC has always dominated with consumer goods: cars, phones, televisions, appliances, computers—all of them use direct current internally. But Edison's plans involved *transmitting* current—sending it over wires from power plants into homes and factories. And when it came to transmitting current, both AC and DC had distinct advantages and disadvantages in the 1880s.

The advantage of DC was that, again, consumer goods like motors ran internally on DC power. If your power supply was also direct current, you could therefore avoid the mess and inefficiency of converting from AC to DC before plugging in. The disadvantage of DC was the huge upfront cost. Given the limitations of transmitting direct current then, Edison needed to build power plants every few blocks—and do so mile after mile. Plus, Edison had to connect the power plants to people's homes with copper wires, and copper was a pricey metal. Edison then made things even harder on himself by insisting that his company bury its wires belowground. For various reasons, he hated seeing wires strung overhead—too ugly, too dangerous, too liable to break. His company began tearing up cobblestone streets instead and laying its wires beneath them. To his credit, Edison often got right down in the trenches with his crews, heaving up the stones and getting smeared with mud. The undertaking was expensive, however, and his crews were allowed to work only at night, to avoid disrupting traffic.

AC power, in contrast, required less upfront investment. To see why, you can think about electricity moving through a wire as akin to water moving through a pipe. Thicker pipes allow for higher water flow, but they're more expensive to make. If you need to deliver a certain amount of water per day, then, and you're forced to use thin pipes, your best option is to increase the water pressure. High pressure, in other words, can make up for the shortcomings of thin pipes.

A similar dynamic is at play with electricity. Thick copper wires can deliver more power, but they're expensive. To get around that, you have to increase the "pressure" in the wires—what scientists call the voltage. (Many people in Edison's day actually used the term "electrical pressure" to mean voltage.) The key point is this: with alternating current, it's trivial to boost the voltage-pressure for transmission. As a result, people could send loads of power down an AC line even when the wires were thin and used little copper. DC was different. At the time, it was hard to increase the electrical pressure with DC and boost the voltage.

The bottom line was that DC power systems needed fat, expensive copper wires, while AC systems didn't. As a bonus, thanks to the higher electrical pressure, AC systems didn't need to have power plants every few blocks; a single plant could serve a whole city. All these factors put Edison's plan to wire cities with DC at a big disadvantage.

Still, alternating current back then did have one major downside—poor equipment. Unlike with DC, no Edisons had invested their time and genius in making good, reliable AC motors, generators, and transmission gear. As a result, Edison believed that his superior machinery—coupled with his glittering public reputation—would overcome the high cost of construction and copper wires and give him a decisive edge in the marketplace. It all might have worked out that way, too, if not for a young Serbian immigrant named Nikola Tesla.

If you like your scientists eccentric, it's hard to beat Tesla. He claimed to speak to Martians on occasion, and would compulsively calculate the volume of any bowl or cup placed in front of him at meals. "I would not touch the hair of other people except, perhaps, at the point of a revolver," he once said, and he'd get physically ill if he saw a peach or pearl. No one knew why. But few in history could match Tesla for sheer intellectual horsepower. Often he didn't even need to test his inventions—they appeared fully formed in his mind, the gears already whirring. While walking in a city park with a friend once, the friend watched Tesla freeze mid-stride. Then his face fell slack, to the point the friend assumed Tesla was having a seizure. In reality, a new type of electric motor had popped into his brain, whole and unbidden. Once Tesla snapped out of it, he sketched the idea in the dirt with a stick, beaming over its elegance. At that point, actually building the machine was superfluous to him. Tesla knew it would work, and it did.

After studying electrical engineering in Europe, the twenty-eight-year-old Tesla traveled to the United States in 1884; he arrived with four cents, a book of poems, and a letter recommending

Electrical whiz and Edison rival Nikola Tesla. (Photograph by Napoleon Sarony.)

him to Edison. ("I know two great men," it said, "and you are one of them; the other is this young man.") Impressed, the thirty-seven-year-old Edison hired Tesla as an engineer, but the two clashed immediately. Some of this tension sprang from scientific differences. Edison favored DC, while Tesla believed the future belonged to AC. Furthermore, Tesla was something of an elitist, and he scorned Edison's greatest gift—his penchant for hard work. In trying to come up with a better lightbulb filament, Edison and his assistants had laboriously tried thousands of different materials, including horsehair, cork, grass, corn silk, cinnamon bark, turnips, ginger, spider silk, and macaroni. This scattershot approach drove Tesla batty. "If Edison had a needle to find in a haystack," he once complained, "he would proceed at once with the diligence of the bee to examine straw after straw until he found the object of his search . . . I was almost a sorry witness of such doings, knowing that a little theory and calculation would have saved him 90 percent of his labor." Why didn't everyone just hallucinate brilliant new ideas like him?

What really produced sparks, however, was the friction of their personalities. Tesla was a neurotic germophobe who dressed in elegant suits. Edison was sloppy and uncouth, with stained shirts and dirty fingernails that disgusted the Serbian. (A reporter once said that Edison "looked like nothing so much as a country store keeper hurrying to fill an order of prunes.") And while it's hard to imagine Tesla ever smiling, Edison loved boneheaded practical jokes. One favorite involved hooking up a battery to a metal sink and turning a crank to build up a big charge. When some dupe touched the sink and leapt back in pain, he'd howl with laughter.

That penchant for jokes, in fact, ultimately destroyed his relationship with Tesla. In the spring of 1885, Edison was at wits' end trying to redesign some DC generators. They were inefficient and prone to breakdowns, and he couldn't see a way around the problems. He told Tesla he'd pay him $50,000—$1.5 million today—if he could fix the flaws. Tesla worked himself to exhaustion, vastly improving the generators' performance. But when he went to claim his bonus, Edison doubled over laughing. "Tesla," he said, "you don't understand our American sense of humor." Edison then claimed—perhaps falsely—that he'd been clowning around the whole time and had no intention of paying such a ridiculous sum. Tesla quit on the spot, unspeakably furious. He was reduced to digging ditches for a while in order to eat, but he refused to work for a liar.

Quitting the job, however, ultimately benefitted Tesla. He soon landed in Pittsburgh with entrepreneur George Westinghouse, who was investing heavily in AC technology. Hiring an unknown like

Tesla was a gamble, but over the next few years, the move paid off handsomely. Tesla eventually earned forty different patents for the Westinghouse company on AC devices, eliminating many of the problems plaguing the technology. To be sure, just like with Edison and the lightbulb, Tesla didn't do everything alone. Other people invented key pieces of equipment, and Tesla rather snobbishly disdained the work of actually implementing his ideas, leaving that to those beneath him. But between Tesla's genius and Westinghouse's business savvy, AC power suddenly looked formidable.

A hiccup in the commodities market soon boosted AC's chances even more. In 1887, some greedy French speculators cornered the world supply of copper, driving its price up to 20 cents per pound ($3 today), twice as much as before. This didn't hurt Westinghouse much, since his company could still use thin wires and just boost the voltage-pressure. Edison, meanwhile, faced ruin. Because his DC system couldn't boost the voltage easily, he needed thick wires to deliver power, and the sudden rise in copper costs threatened his entire vision.

Even worse for Edison, Westinghouse was aggressive. Westinghouse had opened his first AC plant in Buffalo in November 1886. Less than a year later, sixty-eight more plants were open or under construction. AC was proving especially popular in the small towns and suburbs where the vast majority of Americans lived then. The low population density in those places meant that building power plants every few blocks didn't make sense. Westinghouse's scheme proved far cheaper overall.

Pretty soon, Edison was facing checkmate. So, getting desperate, he made the one move he had left. If he couldn't beat AC on merit, then he'd beat it on public relations. He would declare AC a public menace, and use the bully pulpit of his fame to discredit it in people's minds. In short, he'd declare war—what historians now call the War of the Currents.

For Edison this war was only partly about money. Yes, he wanted to fund his beloved research lab, but he'd also made his reputation as an electrical wizard, and the thought of being bested in this arena enraged him and threatened his scientific ego. Edison also still dreamed of revolutionizing America through electrical power—but only if America did things his way. In fact, he'd ousted his company's entire board of directors a few years earlier when they'd challenged his vison of what the electrical industry in America should look like. He'd replaced them with cronies of his, and this sort of groupthink can lead to ethical blind spots—or worse. In sum, Edison found himself in a winner-take-all competition, and losing would threaten not only his bank account but his sense of self; the danger was personal. Psychologists have noted that people in those circumstances are all too willing to trample moral niceties and play dirty, and when he started smearing AC in public, there was no one around to tell him to stop.

Now, Edison's claims about the danger of alternating current did have a modest basis in fact. There's no question that DC at high voltages can be deadly: Lightning, after all, is direct current. But the push-pull, back-and-forth nature of AC does more damage to body tissue, and at a given voltage, AC is more likely to kill (usually by damaging your heart or frying your nerves). Add to that the fact that AC power plants transmitted at much higher voltages in the first place, and things did look scary.

At least to the uninformed. Because while AC would be *transmitted* at high voltages inside power lines, those voltages would be "stepped down" to much safer levels inside people's homes. In demonizing AC, however, Edison always neglected to mention that inconvenient truth. Other things he claimed were outright lies. He told newspapers that, in a house wired for AC, any metal object might kill its inhabitants—doorknobs, railings, light fixtures. As a result, people in AC homes were suddenly afraid to ring buzzers or use house keys. Another bogus claim involved burying wires. Again, Edison's crews buried transmission lines beneath cobblestone streets, whereas Westinghouse's strung the wires overhead, where they might break and shock people. But Edison declared that even if

Westinghouse buried his wires, AC would "come up the manholes" and attack people, like a sewer monster. In Edison's telling, there was no safe level of alternating current.

To be fair, fin de siècle American capitalism was pretty rough-and-tumble, and the claims Edison made, however false, might have been forgivable if he'd stopped there. But Edison soon decided that smears weren't enough. He needed to *show* people the dangers of AC—make them cringe. In short, in a dog-eat-dog world, he decided the best way to get ahead would be to kill some actual dogs.

Edison didn't pioneer the use of electricity to kill animals. That distinction belongs to another man, who was engaged in a battle over the future of the death penalty.

In the 1880s, New York State was seeking a more humane way to execute criminals. The standard method, hanging, had too many bad associations—not only with Southern lynchings but also with the debauchery of public executions in Europe, where drunken revelers gathered to leer at the victim and anatomists brawled over the body afterward. Executioners often botched the job anyway. They'd give prisoners too little rope and leave them dangling in gurgling agony—or too much rope and inadvertently decapitate them when they fell and snapped. Not to mention that prisoners often vomited, soiled themselves, and ejaculated mid-hanging. Not exactly wholesome.

In 1886, New York appointed a three-person committee to devise a better way. First things first, the trio scoured the annals of history and picked out forty possible methods of capital punishment for consideration, including crucifixion, exposure to serpents, boiling in oil, the iron maiden, defenestration, shooting people from cannons, and running the gauntlet. All were rejected as cruel. In the end, support coalesced around two fairly modern methods: lethal injection and electrocution, both of which seemed to kill people gently. In August 1881, for instance, a man named Lemuel Smith had broken into an electrical plant in Buffalo with his friends to touch some poorly grounded equipment, which gave them a pleasant tingling sensation. Later that night, after getting roaring drunk, Smith snuck back in for more jollies and accidentally electrocuted himself. The autopsy showed little internal damage, and from this and similar accidents, doctors concluded that electricity killed people instantly and without pain.

Still, two committee members voiced support for lethal injection. That's when the third member, Buffalo dentist Alfred Southwick, who supported electrocution, took matters into his own hands. The city of Buffalo had recently started paying 25¢ for every stray dog turned into the pound. Local urchins took full advantage, and the pound's cages were soon overstuffed with mutts—far more than attendants could care for. Southwick stepped in and offered to help them cull. He built a wooden cage with a zinc floor connected to a local power line. He then filled the box with an inch of water and placed a terrier inside. It wore a metal muzzle, also connected to the power line. When everything was ready, Southwick flipped a lever and completed the circuit. The terrier slumped over dead. Further trials dispatched twenty-seven more dogs, none of which yelped or bucked or showed any signs of suffering.

These tests convinced Southwick that electrocution was the perfect mode of death. To bolster his case, he then sat down in November 1887 and wrote a letter to the most famous electrician in the world. He wanted Thomas Edison's endorsement in supporting this quick, easy method of execution.

Edison rebuffed him. He told Southwick he found the death penalty barbaric and opposed capital punishment on humanitarian grounds. (As Edison once put it, "There are wonderful possibilities in each human soul, and I cannot endorse a method of punishment which destroys the last chance of usefulness.") In short, he would never support Southwick's cause.

However chagrined, Southwick wrote Edison back in December. Nations have executed criminals since the beginning of time, he argued. Given that reality, shouldn't we strive to minimize suffering and find more humane ways of putting people to death?

Southwick no doubt expected another tongue-lashing. But Edison's answer surprised him. Although Southwick couldn't have known it, the exchange took place during the middle of Westinghouse's big expansion of AC power plants. DC technology was on the ropes, and Edison's genius was taking a pummeling. While Edison didn't allude to these affairs in his response, his answer looks suspicious in light of them. He would certainly abolish the death penalty if he could, he wrote. But until that day came, nations should strive to adopt "the most humane method available," and electrocution fit the bill. He then helpfully added that, while several different generators could kill, "the most effective of these are known as 'alternating current' machines, manufactured principally in this country by Mr. George Westinghouse."

Thrilled, Southwick took Edison's letter to the committee members who'd been leaning toward morphine injections. This changed their minds. If Thomas Edison supported electricity, well, that was good enough for them. In early June 1888, they publicly recommended electrocution for New York State.

Despite Edison's none-too-subtle hint, the committee didn't specify whether to use alternating or direct current for executions, leaving that choice to the future. The next day, however, an Edison partisan published an incendiary letter in a newspaper to influence their decision. He denounced AC as a "damnable" technology, and added that hanging AC lines above New York streets "is as dangerous as burning a candle in a [gun]powder factory."

Seeing this letter, and sensing trouble, George Westinghouse wrote Edison a few days later with a peace offering: "I believe there has been a systematic attempt on the part of some people to do a great deal of mischief," he said, and exacerbate the conflict between them. Let's put an end to it. He also extended an offer. Years earlier, before Edison viewed him as a threat, Westinghouse had toured Edison's labs in Menlo Park, New Jersey. Westinghouse now proposed returning the favor and letting Edison tour his headquarters in Pittsburgh, to establish "harmonious relations."

Edison spurned the offer. Too busy to travel, he claimed.

Remarkably, though, Edison did find time to hatch another plot against Westinghouse. The newspaper letter had kicked up a hornet's nest among engineers about the merits of AC versus DC, and a reporter called Edison in mid-June for comments. Edison invited him out to his lab for a demonstration instead. The reporter arrived to find a dog with a rope around its neck. It was standing on a tin sheet, which was connected to a generator. A nearby water dish was connected to the generator as well. When the dog leaned down to drink, Edison explained, it would complete the circuit and kill itself.

The dog, however, refused to cooperate. Sensing something amiss, it wouldn't drink on its own; when Edison's assistants yanked its head down with the rope, the dog snapped it and ran away. The assistants replaced both rope and dog and resumed the tug-of-war. Finally, after one hard pull, the dog slipped. Its paw splashed into the water dish, and 1500 volts coursed through its heart and brain. After a single yelp, it dropped dead. The reporter was impressed and wrote up a story. In it, he dutifully noted Edison's main point—that they'd used alternating current.

Things quickly got worse from there. The author of the original newspaper letter was an electrician named Harold Brown, who more or less worshipped Edison. But his diatribe had been denounced by several engineers, who maintained that he had too little evidence to support his claims about the dangers of AC. So despite having never met Edison, Brown wrote to the Wizard of Menlo Park and asked whether he could use the labs there to generate more evidence—by electrocuting more dogs.

To Brown's surprise, Edison agreed. In truth, opening up his lab to strangers wasn't unusual for Edison, who could be quite generous sometimes. In this case he even loaned Brown his top assistant to help out. What was unusual here was the conditions Edison put on the work. Normally Edison encouraged collegiality and the open exchange of ideas—the scientific ideal. But he told Brown to keep mum about these experiments. He also restricted Brown to working at night, so that people wouldn't hear the howls.

As in Buffalo, someone posted a sign near Edison's lab offering a quarter apiece for stray dogs, and local ruffians once again came through. Brown planned to electrocute the mutts systematically, but in reality the work was haphazard. The dogs differed wildly in size—setters, terriers, Saint Bernards, bulldogs—and he zapped them with both DC and AC at anywhere from 300 to 1400 volts. The results were nevertheless consistent—an uninterrupted litany of suffering. The dogs jumped and yelped and whimpered in pain, and those that weren't stunned by the shocks made "violent effort[s] to escape," he noted. One started bleeding from its eyes.

After a month of this, Brown felt confident enough to arrange for the demonstration above, where he tormented a Newfoundland mix at Columbia. The newspaper coverage was outraged, and any normal man would have slunk away in shame. Brown, in contrast, staged another demo a few days later, killing three more dogs with alternating current and allowing doctors to dissect them afterward. All in all, he reported to Edison's assistant, the experiments were a "fine exhibit" about the dangers of AC.

Others disagreed. Not only was Brown being cruel, they argued, but his experiments proved nothing. In shocking some of the dogs with DC first, he'd battered and weakened them, making it impossible to determine how much each type of current had contributed to their deaths. Furthermore, dogs were small animals. If humans were shocked with AC, there's no guarantee they'd react the same way.

In response to these criticisms, Brown held yet another demonstration in December 1888, at Edison's lab. This time he fried big animals, and used AC alone to do so. He started with a 124-pound calf, attaching an electrode between its eyes; 770 volts dropped it. A second, 145-pound calf succumbed to 750 volts. Then, to quell

Depiction of the "experiments" to electrocute horses. Notice the doghouses in the background, where more animal victims await.

all doubts, Brown and Edison's assistant wired up a 1200-pound horse they'd acquired for $15, attaching the electrodes to two different hooves so that the current coursed through its heart. Edison had previously promised reporters that AC killed beasts in one ten-thousandth of a second. In reality, the nag survived five seconds at 600 volts, then fifteen more seconds at the same voltage. Finally a twenty-five-second pulse at 700 volts dispatched it. Edison paid $5 to cart the carcass away.

In doing these experiments, Brown succeeded in his main goal: terrifying people about AC. Still, Edison's team recognized that torturing dogs and horses wouldn't exactly endear them to the public. In his private notebook, Edison's chief electrician cringed at how much the beasts had suffered. In a magazine story printed shortly afterward, however, he nevertheless insisted that their deaths had been "instantaneous and painless."

Not everyone bought this propaganda. One critic dismissed Brown as a "lizard-blooded scientific promoter of murder." Edison caught heat as well: Westinghouse more or less publicly accused him of hiring Brown to do his dirty work. Laughably, Edison denied this, claiming that Brown worked completely independently—despite the fact that Edison had loaned him lab space, equipment, and assistants.

In response to the accusations against him, Brown challenged Westinghouse's manhood and took out newspaper ads proposing that they engage in a duel—an electricity duel. If Westinghouse was so sure that AC was safe, Brown said, let's wire both of us to generators, Brown to DC and Westinghouse to AC. They'd start with a zap at 100 volts, and move up in increments of 50 until someone cried uncle—or dropped dead. "To the regret of many in the industry," one historian noted, "the duel never took place."

Eventually, Edison's team would kill forty-four dogs, six calves, and two horses in their quest to discredit alternating current. Edison even sought out circus elephants to kill,* and was crushed when the plans fell through. But none of these deaths did any good—Westinghouse continued to crush Edison in the marketplace. By the end of 1888, Edison's company was building and selling enough equipment to power 44,000 lightbulbs per year. Westinghouse sold

*Elephant executions were surprisingly common. As wild beasts, elephants resented being locked in tiny cages at zoos or prodded into doing tricks for circuses. Some trainers were also needlessly cruel: one got drunk once and fed his elephant a lit cigarette. Unsurprisingly, these abused elephants sometimes lashed out and killed people, at which point they'd be executed. One scholar dug up thirty-six separate cases of pachyderm capital punishment. This included Topsy the elephant, who was killed in 1903 via electrocution. Understandably, given that Edison wanted to kill an elephant himself, and given that he did dispatch many other animals via electricity, many people today believe that Edison personally had a hand in Topsy's demise. That isn't true—the War of the Currents was long over by 1903. But Edison's film company did record the execution, which makes for some brutal viewing.

enough equipment to power 48,000 lightbulbs in October 1888 alone.

Edison had one hope left. To salvage direct current, he'd need to make the connection between AC and death so stark that no one could deny it. He'd have to dispatch a human being.

On the morning of March 29, 1889, in Buffalo, New York, a dipsomaniac fruitmonger named William Kemmler beat his wife Tillie to death with the blunt end of a hatchet. She'd been flirting with another man, he claimed, and deserved everything she got. After wiping the blood off his hands, the twenty-eight-year-old Kemmler strolled down the street to a bar for his morning eye-opener, where police arrested him. Even Kemmler's lawyer called him "monstrous," and Kemmler was inclined to agree: "I'm ready to take the rope," he said. He didn't realize that New York State had outlawed hanging, and that he was now slated to become the first person in history to die in the electric chair.

The chair would be located at Auburn State Prison near Syracuse. Officials there—dazzled by the name of Edison—had enlisted his lackey Harold Brown to help construct it, and Brown naturally recommended they use Westinghouse generators. When Westinghouse refused to sell any to the prison, Brown paid a third party to locate some used ones and scratched off the serial numbers so no one could trace them. Edison's minions then trumpeted the selection of Westinghouse equipment in the press. (Later, a stack of letters stolen from Brown's desk provided strong circumstantial evidence that Edison had paid Brown $5,000 [$150,000 today] to build the chair. As for how the letters disappeared from Brown's desk, no one knows. But some historians believe that Westinghouse—who could play just as dirty as Edison—arranged for the burglary.)

To fight back, Westinghouse bribed members of the New York legislature to abolish the death penalty. When that tactic failed, he went to the courts. Given Brown's torture of dogs and horses, there were serious questions about whether the electric chair would be cruel and unusual punishment. In fact, when Kemmler's high-powered lawyer, Bourke Cockran, was asked why he took the case, he explained that his wife had heard about those poor dogs and couldn't stand the thought of someone doing that to their family pooch. In reality, Westinghouse was secretly paying Cockran $100,000 ($3 million today); otherwise, Kemmler never could have afforded him. But Cockran did raise legitimate fears about electrocution.

Alas, Cockran's objections never had a chance. At a hearing about the cruel-and-unusual question, the state's lawyers called in the smartest and most honorable witness they could think of to help decide the matter, Thomas Edison. Despite cheerily admitting that he knew nothing about anatomy or physiology, Edison swore that Kemmler would die instantly and painlessly in the chair—provided they used alternating current. In private, he and Brown had even taken to calling AC "the executioner's current."

(This wasn't the Edison team's only attempt to bend the English language against his rival. The word *electrocute* hadn't caught on yet, so journals and newspapers solicited proposals from readers about what to call death via electricity. The public responded in droves, suggesting *electricize, voltacuss, blitzentod, electrostrike,* and *electrothanasia,* among other words. Edison's lawyer's suggestion was more pointed: Kemmler would be *westinghoused.*)

Thanks to Edison, Kemmler lost his suit to ban the electric chair. Two days later, on October 11, 1889, the world got a preview of what awaited him. Just after noon that day, an electrical repairman got tangled in the spider web of power lines above a street in downtown Manhattan and accidentally touched a live wire. He likely died within seconds, but because he was trapped in the wires, electricity continued to course through his body. Like some biblical demon, blue flames erupted from his mouth, and sparks burst forth from his shoes. Thousands of people gathered below to gawk and scream, despite occasional sprays of blood. But the incident apparently didn't

The notorious first electric chair, at Auburn State Prison in New York. (Courtesy of the Library of Congress.)

shake anyone's faith that Kemmler's death would not be cruel. Thomas Edison had promised, after all.

Kemmler was finally scheduled to die just after dawn on August 6, 1890. He entered the execution chamber looking preternaturally calm, and spoke a few soft words to the gathered witnesses and reporters. He'd recently gotten his hair cut for his big day, but prison guards ruined it by shaving a tonsure and attaching an electrode to his skull. They also slit his shirt and attached another electrode to his spine. Kemmler then took a seat. (Aside from the obvious, the chair was reportedly quite comfy.) When one of the guards began fumbling with the leather straps that would hold his arms down, Kemmler cooed, "Don't get excited, Joe. I want you to make a good job of this." As a last step, the warden fitted a leather mask on his face. Then the warden rapped on a nearby door, the signal for the electrician in the next room to throw the switch.

As the current bit into him, Kemmler snapped upright. His mouth curled into a mockery of a grin, and one of his fingernails dug so deep into his palm that he started bleeding. Seventeen seconds later, it was over. The electrician cut the current, and Kemmler slumped over like so many dogs before him. Doctors on hand pressed his face with their fingers and pointed out the mottled, red-and-white after-impressions—an unmistakable sign of death, they said. Among the witnesses was Alfred Southwick, the Buffalo dentist who'd executed the dogs from the pound. "This is the culmination of ten years' work and study," he announced. "We live in a higher civilization today."

The only problem was, Kemmler wasn't dead. His palm was still bleeding, and one of the witnesses noticed that the spurts of blood were rhythmic—a sign of a heartbeat. "Great God, he's alive!" someone shouted. As if on cue, Kemmler moaned like a wounded sow and convulsed, spitting up purple foam through his mask.

The room was pandemonium. "Turn on the current!" someone screamed. Unfortunately, no one had considered the need for a second pulse, and it took the electricians a few minutes to get the generator going again. In the meantime, Kemmler continued to groan and quake.

At long last, the current snapped on again. In the chaos, no one remembered how long the second pulse lasted; estimates ranged from sixty seconds to four-and-a-half minutes. But it was enough to kill Kemmler,* and then some. An odor of burning hair and fried skin filled the room. One witness vomited. Another fainted. A third wept.

During the autopsy, Kemmler's body was so stiff that it remained in a seated position on the table. Doctors discovered that the electrodes had burned through his back into his spine, and that most of

*As historians have noted, Kemmler probably doomed himself by being too calm and collected in the face of death. If he'd only panicked a bit—like the guards and witnesses on hand—the sweat on his skin probably would have conducted the electricity into his body more efficiently and helped kill him right away.

his brain had been carbonized into black embers. The physicians nevertheless had to wait three hours to declare Kemmler dead. The legal definition of death back then was the point at which the body could no longer produce its own heat. Kemmler's body was so hot from being westinghoused that it didn't cool down until mid-morning.

As part of the price of admission, the newspaper reporters on hand had promised not to reveal anything about the death except the bare facts. But to hell with that—this was the hottest scoop of the year, and the headlines were screaming. Southwick gamely tried to claim that things had gone well. The death was so mild, he said, "A party of ladies could have been in that room." Other witnesses were more honest. "I will see that bound figure and hear those sounds until my dying day," said one. Westinghouse did not witness the death, but he summed things up aptly: "They could have done better with an axe."

Thomas Edison acknowledged there were bugs to fix, but predicted that the next execution "will be accomplished instantly and without the scene at Auburn today." He was not a cruel man—he didn't relish Kemmler's suffering. But all was fair in war. Besides, what could you expect from a technology as dangerous as alternating current?

It might be tempting to excuse Edison and Brown's behavior on the grounds that theirs was a different era, a time when society simply didn't treat animals well. But many people back then did protest cruel scientific research, and had been doing so long before Edison's day.

Voltaire sneered at "barbarians . . . who nail [a dog] on a table and dissect it alive." Samuel Johnson seconded that, and added, "He surely buys knowledge dear, who learns . . . at the expense of his humanity." Anatomist John Hunter was a frequent target of such attacks, since he often practiced new surgical techniques on shrieking dogs and pigs; he also did things like inject vinegar into a pregnant dog's veins just to see if it would abort. (It did.) A few entomologists even protested sticking live insects on pins, since they squirmed in agony afterward, sometimes for days. These protests weren't isolated voices, either. The powerful Hearst newspaper chain loudly condemned "vivisectionists" for abusing animals. Edison's latter-day defenders can't plead ignorance for him.

Conditions have clearly improved since Edison's time, but experiments involving animals remain controversial today, even among some scientists. This is partly due to the sheer number of animals that die. Medical research exploded in the second half of the twentieth century, and by the year 2000, American scientists alone were going through half a billion mice, rats, and birds per year, plus dogs, cats, and monkeys on top of that. The scale is staggering.

The obvious rejoinder is that animal research saves human lives, through the development of drugs and other treatments. While that's certainly true, there are caveats. However useful animal research was in the past, it often falls short of expectations nowadays. One survey of twenty-six known human carcinogens found that fewer than half also caused cancer in rodents; for predictive value, you might as well flip a coin. Things are even worse with new medicines. In 2007 the U.S. Secretary of Health and Human Services—not exactly an obscure source—admitted that "nine out of ten experimental drugs fail in clinical studies because we cannot accurately predict how they will behave in people based on laboratory and animal studies." Such failures are in fact so common they're almost cliché. How many times have we heard about some amazing therapy that miraculously stops cancer or heart disease or signs of Alzheimer's in mice—only to watch it flop in human beings?

Perhaps this shouldn't surprise us. Evolutionarily, rodents and humans diverged 70 million years ago, back when dinosaurs still ruled the Earth, and we have notably different physiologies. Penicillin is actually fatal to that proverbial lab animal, the guinea pig; had scientists initially tested this drug on them, it never would have made it to market. Even our close evolutionary cousins have different biologies: HIV devastates the human immune system, but is a harmless, slow-growing virus in chimpanzees. Given these facts, some critics of animal testing have been scathing. One called animal research "an internally self-consistent universe with little contact with medical reality."

To be sure, animal research does still produce cures. If nothing else, it helps screen out poisonous drugs before they're ever tried in humans, which is no small thing. But in the past few decades there's been a movement to cut back on the number of animals used in labs and find alternatives. Possible alternatives include running tests on human organs grown in dishes (organoids) or using computer programs to estimate the efficacy of new chemicals by comparing them to known compounds. Some animals have also won low-level legal rights. The U.S. government no longer supports biomedical research on chimpanzees, and the requirements for using monkeys in general are strict. Similarly, the EPA recently announced that it will phase out toxicity tests with mammals by 2035 and severely curtail tests on birds. (Tests on amphibians and fish will continue.) Perhaps most surprising, the sheer brilliance of octopuses* has convinced some international groups to require scientists to seek special permission to experiment on them. That's especially significant since octopuses are invertebrates, animals we usually exempt from our moral codes.

In all, life is vastly better for research animals today compared to the 1880s. But reports of abuse still pop up in labs around the world, and outré experiments (e.g., monkey head transplants) have not ceased. The howls of Edison's dogs continue to echo today.

---

*Among other tricks, octopuses can juggle objects and open jars—without having been taught how. Or consider Otto, an octopus at an aquarium in Germany. Apparently Otto resented the lights that were shining into his tank at night. So he learned to haul himself up to the tank's edge and squirt water at the lights, shorting them out. He repeated this process three nights in a row, baffling the aquarium staff who couldn't understand why the circuits kept burning out. They finally took to sleeping on the floor at night before they caught him.

Ultimately, not even the torment of William Kemmler could negate the advantages of alternating current. Before the 1893 World's Fair in Chicago, the General Electric company submitted a bid to light the grounds with Edison's DC equipment for $554,000 ($16 million today). Westinghouse underbid them by $155,000 and won the contract instead. After that, the gap in quality and cost only widened. By 1896, an AC plant near Niagara Falls was powering Buffalo from the astonishing distance of twenty miles, a span that direct current never could have matched.

Not long after the Niagara plant opened, Edison conceded defeat* in the War of the Currents. Few people in history can match his record of innovations, but his beloved direct current played almost no role in the twentieth-century revolution in cheap electric power.

Some historians have argued that Edison's defeat wasn't inevitable. Had he recognized the downsides of DC earlier, they argue, and switched to AC instead, his prestige alone would have been

---

*Although the internet likes to paint Edison as the villain in Tesla's life, George Westinghouse was the one who really screwed him over. Westinghouse signed Tesla to a generous royalty contract in the late 1880s; it paid Tesla $2.50 for every one horsepower generated by his equipment. Given Westinghouse's fantastic rate of expansion, that would have amounted to $12 million by 1893 ($323 million today), a sum that would have bankrupted the firm. So Westinghouse begged Tesla to tear up their contract. "Your decision determines the fate of the Westinghouse company," he said. Incredibly, Tesla did as he was asked. Unlike Edison, Westinghouse had believed in him, and Tesla felt duty-bound to help. So they voided the contract.

Sadly, Westinghouse was not so generous to Tesla in return. Many years later, when the Westinghouse Company was obscenely profitable, Tesla returned to his benefactor, hat in hand, and asked for some of the money back. Westinghouse refused, and Tesla ended his life more or less indigent, unable to afford even the rent at the hotel in New York where he lived. For more on the sad end of Tesla's life, see episode 18 of the *Disappearing Spoon* podcast at samkean.com/podcast—a tale that, believe it or not, involves Donald Trump.

enough to win out in the marketplace. Perhaps. But without Tesla's patents, he was at a big disadvantage, and Edison was nothing if not stubborn. The real shame was that he didn't bow out with grace, and spare those horses, calves, and dogs the pain and indignity of electrocution. Moreover, while William Kemmler would have been executed anyway, Edison helped put him through one of the most gruesome deaths in the annals of jurisprudence. Tellingly, in later interviews and memoirs, Edison omitted all mention of torturing animals and his role in developing the electric chair.

However heated it was, Edison's spat with Westinghouse and Tesla was just one of many scientific rivalries in history. In fact, another nasty feud between American scientists was also peaking in the late 1800s, and once again, animals were caught in the crossfire. Luckily, the spat between Edward Drinker Cope and Othniel Charles Marsh involved animals who were long past suffering: both men were paleontologists, fighting over fossilized dinosaurs. And unlike the destructive War of the Currents, the Bone Wars not only pushed their field forward but proved one of the most delightfully catty episodes in the history of science.

6

SABOTAGE: THE BONE WARS

Edward Drinker Cope was ecstatic. He'd just scooped his nemesis, Othniel Charles Marsh, and had done so in the most humiliating way possible.

It was August 1872, and teams led by Cope and Marsh were both digging for fossils in southwest Wyoming. Each group was heavily armed and trying to avoid contact with the other, but there was always a dusty wagon track here, or some abandoned tools there, to remind them of the enemy. One day Cope's curiosity got the better of him, and he spent a few hours spying on Marsh's team from afar, concealing himself as they hacked at rocks in the distance.

When they packed up and left, Cope snuck down to investigate. To his delight, he found an overlooked skull fragment; several teeth were scattered nearby as well. In fact, the unusual combination of skull and teeth suggested a brand-new species of dinosaur. That it had slipped through Marsh's fingers no doubt added a bounce to Cope's step as he pocketed the bones and strolled back to camp.

He had no idea the joke was on him. Aware of his spying, Marsh's diggers had "salted" the site with the skull and teeth, which belonged to different species. They were hoping to trick Cope into a public blunder, and he walked right into the trap. He published a paper about his "discovery" soon afterward, only to have to retract it later.

Rivalries are funny things. They waste time and energy. They stir up mean instincts and consume us with petty emotions. Yet in doing so, rivalries also push people to greatness. In their fury to one-up each other, Marsh and Cope discovered hundreds of new dinosaurs and other species and filled whole museums with specimens. Their work also transformed dinosaurs from an obscure taxon of lizard into some of the most famous animals of all time. Like a lovely phoenix, an entirely new understanding of Earth's history, and the place of human beings within it, arose from the ashes of their hatred.

Oddly enough, Cope and Marsh began as friends, despite strikingly different temperaments.

Marsh plodded. He drifted through his youth on a farm east of Niagara Falls, hunting and fishing, and might have gone right on drifting through his whole life if not for his uncle George Peabody. For whatever reason, the wealthy financier took a shine to the lad and paid for his enrollment at the prestigious Phillips Exeter Academy in New Hampshire. (Marsh entered at age twenty; classmates called him "Daddy.") School awakened in Marsh an unexpected passion for natural history, and Uncle George dutifully sent his nephew to Yale afterward. There, Marsh collected so many minerals and fossils in the attic of his boardinghouse that his surprisingly indulgent landlady, who lived beneath, had to reinforce her ceiling to keep the beams from buckling.

Marsh—who had a pinched face and beady eyes—longed to start a family, but he was always awkward around women. (He once called a potential love interest the "prettiest little vertebrate" he'd ever laid eyes on.) Resigned to a bachelor's life, he traveled to Europe after graduating Yale in 1860, and spent several years studying at different museums and universities on Uncle George's dime.

Cope, in contrast, raced through life—the hare to Marsh's tortoise. Cope grew up outside Philadelphia, and was considered a child prodigy in natural history. While working on a farm one summer at age thirteen, he seized a two-foot-long snake by the neck and blithely hauled the hissing, whipping serpent back to his host family. They panicked, screaming that it was poisonous. But in between its attempts to bite him, Cope calmly examined the snake's teeth and explained why they were wrong—it lacked the right fangs to inject venom. So, he said, no need to worry.

Hotheaded paleontologist Edward Drinker Cope in his office. (More pictures are available at samkean.com)

By age twenty-one, Cope—who had a devilish grin and flamboyant mustache—had published thirty-one scientific papers, an impressive start to his career. At the same time, he was also developing a reputation as hotheaded. Although born a Quaker and raised a pacifist, Cope was a brawler by nature; a friend once described his approach to life as "war at whatever it cost." He clashed most often with his father, a merchant who'd bought his son a tract of land and was grooming the boy to go into farming; they quarreled constantly over his future. Cope liked sparring with other scientists, too. He once exchanged blows with a colleague in the hallway outside a scientific meeting, leaving them both with black eyes. The man was Cope's best friend.

In 1861, Cope moved to Washington, D.C., to study at the Smithsonian Institution. Alas, he was something of a lothario and got tangled up in a messy love affair there. Because most of his letters from this period have gone missing (or were destroyed), the details remain mysterious. Was his paramour a charwomen, an heiress, a Capulet to his Montague? No one knows. Regardless, Cope's father sent him overseas to disentangle him from Madame X, a trip that also spared him the possibility of being drafted into the Union army.

As two young American naturalists abroad, Cope and Marsh naturally ran across each other in Europe, meeting in Berlin in 1863. True to form, Marsh, thirty-two, had been patiently studying there for months, while Cope, twenty-three, was blowing through town on a frantic tour, popping in and out of different museums. Marsh later painted Cope as borderline mad in Berlin—an unstable Hamlet still pining over his lost love. Marsh nevertheless took a liking to his younger colleague, and they began swapping letters every few months. After returning to the United States, Cope even named a new type of amphibian after Marsh, and Marsh returned the honor and named an aquatic reptile after Cope.

Still, it didn't take long for their relationship to fray. The first spat involved some dinosaur pits in New Jersey. Dinosaurs were first recognized as something unique in England in 1817, and amateur fossil-hunters like Mary Anning were responsible for several early discoveries there. But no one knew dinosaurs existed in North America until 1858, when naturalist Joseph Leidy recovered the bones of a duck-billed dinosaur (*Hadrosaurus*) from a quarry in New Jersey. (Typically, the quarry workers noticed the bones initially, during routine digs; the quarry owners then alerted scientists, who took over after that.) With Leidy's blessing, Cope began working in the quarries in 1866 and excavated a carnivorous dinosaur (now called *Dryptosaurus*). The discovery so excited Cope that, to his father's frustration, he quit his teaching job the next year and moved his new wife and daughter closer to the pits to dig full-time. To capitalize on their discoveries, Leidy and Cope hired a sculptor to mount a twenty-six-foot replica of the *Hadrosaurus* at a museum in Philadelphia—the first mounted dinosaur skeleton in history. It was a brilliant fusion of art and science,* and word about it soon reached Marsh in New Haven.

Like Cope, Marsh had done well for himself lately. He'd been pressing his Uncle George to establish a natural history museum at Yale, and in a bit of genteel extortion, he'd let Yale know that he expected to be rewarded for brokering the deal. Peabody finally coughed up $150,000 ($2.6 million today) and in exchange, Yale appointed Marsh trustee of the museum and named him a professor of paleontology, the first such post in North America.

Professionally, then, Marsh had reached the pinnacle of U.S. fossil-hunting. Scientifically, though, Cope and his New Jersey specimens were hogging all the glory. So, Marsh wrote to Cope to ask if he could visit the quarries. Cope agreed, and in March 1868 the two spent a happy week tramping about in the rain and snow, digging and exploring. At the end of which, Marsh thanked Cope for his generosity, left for the train station—and quickly doubled back to the pits. There, he bribed the quarry owners to cut Cope out

---

*While Leidy and Cope's was the first complete skeleton ever mounted, it was not the first reconstructed model of a dinosaur in history. Based on some pitiful bone fragments, paleontologists in England had made some wildly speculative sculptures of several dinosaurs for a park in London in the 1850s. They were such a big hit that officials in New York planned to erect a similar set of monsters in Central Park—and would have, except that none other than Boss Tweed came along and destroyed them, then ran the sculptor out of town. For more on this wild story, see episode 6 at samkean.com/podcast.

Incidentally, Marsh once mounted a skeleton of one creature (a rhino-like mammal called an "uintathere") by using papier-mâché to make molds of its bones. For the papier, he got hold of the thickest, sturdiest material he could find: shredded U.S. greenbacks. As a result this beast was, perhaps literally, a million-dollar skeleton.

and start routing the best fossils to him instead. After that, all the choice bits ended up at Yale.

Cope didn't learn of this duplicity until later, but by that point he and Marsh had already had a falling-out over another incident. A few years prior, railroad workers hacking through some shale out in Kansas had uncovered a spectacular plesiosaur, an extinct aquatic reptile. The skeleton ended up with Cope, who named the beast *Elasmosaurus*. The name means "thin-plate reptile" or, more colorfully, "ribbon reptile," after its extraordinary long tail, which stretched dozens of feet. Cope then mounted the skeleton for display at a museum in Philadelphia and dashed off a paper about its anatomy.

Marsh once again visited Cope to view the skeleton, and was once again boiling with jealousy. Upon taking a closer look, however, his frown flipped upside down: He'd spotted a blunder. In his haste, Cope had inverted the vertebrae. That is, he'd mistaken the top of the spine for the bottom of the spine, and had consequently mounted the skull on the beast's posterior. The ribbon-reptile didn't have an immensely long tail at all; it had an immensely long neck.

Marsh later swore he'd been gentle in pointing out the mistake. Cope maintained he'd been "caustic" and cruel. Regardless, the two men fell to quarreling over the spine's orientation. To arbitrate the matter, they called in Leidy, who also worked at the museum. After looking things over, Leidy plucked off the skull, took a long walk down to the tip of the "tail," and reattached it there.

Cope was mortified. However prolific, he was still a young scientist, and a high-profile mistake like this could derail his career. He began buying up and destroying every last issue of the journal with the *Elasmosaurus* paper in it, even asking colleagues to mail him their copies, at his expense. (He later printed a new issue with the mistake corrected.) Marsh dutifully did as Cope asked and sent his journal in—then bought two more copies on the sly, which he kept for the rest of his life. He regarded the whole incident as hilarious. Cope, meanwhile, was furious, and never forgave Marsh for exposing him.

Even if Marsh had never embarrassed Cope, the two men's temperaments would have driven them apart eventually. Cope was speedy, Marsh slow. Cope was charming, Marsh guarded. And while Marsh believed wholeheartedly in Charles Darwin's new theory of evolution, becoming an early champion in the United States, Cope sympathized with creationists and struggled to accept evolution as a fact. Even when he did, he reserved a role for God in the process, a notion Marsh sneered at.

Still, despite their mutual distaste, their relationship probably wouldn't have deteriorated into outright hatred if not for a change of scenery. While ensconced in comfortable museums out East, their antipathy remained civil. Once they shifted to the Wild West, Cope's war at whatever cost became inevitable.

Millions upon millions of years ago the interior of North America was a gigantic inland sea, an American Mediterranean. Untold numbers of creatures died and were buried in its depths and on its shores, and erosion and tectonic uplift eventually exposed their remains. The result was one of the richest fossil beds in history. Through the mid-1800s, fossils were so abundant in some places out West—just lying on the ground, like primordial picnic remains—that a shepherd in Wyoming once built an entire home out of ancient bones, an osteological log cabin.

Word of this bonanza began to trickle back East after the Civil War, and in 1870 Marsh organized a fossil-hunting expedition out there, paid for in part by a bequest from his Uncle George. His primary field hands were a dozen whippersnappers from Yale, but the U.S. Army provided critical support. In the 1870s, it was actually easier to travel from the East Coast to Europe than to many places west of the Mississippi, and Marsh's crew depended heavily on the army and its frontier forts for supplies. Moreover, given the U.S. government's efforts to push Indian tribes off their lands (if not exterminate them), Marsh's crew likely would have been ambushed and killed without military muscle. In fact, at Marsh's first stop, a fort in Nebraska, he met an antelope hunter who'd staggered in the day before with an arrow sticking out of him.

Overall, seventy people made up Marsh's initial party, including army escorts and a few Pawnee guides. Everyone carried a bowie knife, carbine, and six-shooter. The most notable among the

Brooding paleontologist Othniel Charles Marsh with Chief Red Cloud, who called Marsh "the best white man I ever saw." (Courtesy of Yale University.)

entourage was William Cody, later known as Buffalo Bill, the Wild West showman. (Not yet famous then, Bill worked for the army as a scout.) As the party headed off, Bill listened to Marsh lecture about the giant thunder-lizards of yore, and how all the dust-choked land around them had once been underwater. Bill just nodded, smiling inwardly and pretending to play along. He'd told some tall tales in his day, but he had to hand it to Marsh: he'd never told a whopper as big as that!

Buffalo Bill peeled off the second day, declining to help dig up fossils. But the soldiers accompanying Marsh eagerly pitched in. (The Pawnee were more reluctant, until Marsh showed them some ancient horse fossils, which delighted them and changed their minds.) When searching the bluffs for bones, the diggers looked not only for shape but texture—bone was smoother and shinier than rock, and often had a telltale spongy interior. Once located, the fossils were hacked out with chisels, knives, spades, or picks, whatever it took. The diggers also spent agonizing hours crawling around on their knees, noses in the dirt, searching for bone fragments or teeth that had worked themselves loose naturally. Delicate structures were wrapped in cotton or frontier newspapers and tucked into cigar boxes or tin cans for the trip back East. Hulking femurs—some of which weighed a quarter-ton—might be encased in strips of burlap dipped in plaster of Paris, the same basic method doctors used to make casts for fractures.

To get from site to site, the troops marched for up to fourteen hours at a stretch, in temperatures approaching 120°F. Food was decently plentiful—buffalo steaks, stewed jackrabbit, canned vegetables and fruits—but water so scarce that they sometimes had to fill their hats during thunderstorms and slurp it down. Bears and coyotes hounded them, and rats and salamanders swarmed their tents at night. But for Marsh, the hardships couldn't undermine the thrill of collecting. In addition to dinosaurs, his crew unearthed mastodons, ancient camels and rhinoceroses, and several different species of extinct horse. In fact, when they reached Utah, Brigham Young himself interrogated

Marsh about the horse remains. This encounter baffled the naturalist, until Young revealed the reason for his interest. According to Mormon theology, horses originated in the Americas, not Eurasia, and he was looking for evidence to this effect. No naturalist back then would have supported the notion, but Marsh's work would eventually prove Young right. (The Yale field hands, meanwhile, were more interested in interrogating Young's twenty-two daughters, whom they flirted with in a box at the local theater.)

By trip's end, in December, Marsh had sent whole train cars full of fossils back to Yale. But his most celebrated find came during his very last hour in the field, in Kansas. While poking around some rocks off a trail, he noticed half a bone lying on the ground. It was six inches long and hollow, like a thick drinking straw. He pegged it as part of a hand bone, equivalent to a section of the pinkie. But what species, he didn't know.

Alas, the light that day was fading, and Marsh had no time to hunt for the bone's other half. All he could do was carve a cross in the rock face nearby, to mark the spot, and return next season.

Marsh spent the winter brooding over the fragment. Based on its distinct shape, he concluded that it belonged to a pterodactyl. The only problem was, all known species of pterodactyls then had tiny wingspans, hawk-sized or smaller. If this bone really did correspond to its pinkie, the beast must have been a goliath, with wings at least twenty feet across—a "dragon," Marsh remembered thinking.

It was exactly the kind of discovery that would win him glory—provided he was correct about the dragon's size. But what if the bone's other "half" was actually much smaller, or what if this wasn't part of its pinkie at all? Uncharacteristically, Marsh threw off caution and rushed something into print. He then spent the next few months fretting. Would this be his skull-on-the-tail moment, the cudgel by which Cope would exact revenge?

Marsh made the Kansas site his first stop the next spring. As soon as the crew's tents were pitched, he raced off and found the cross he'd carved into the chalk. After a few minutes of hunting, he found the bone's other half, as well as several other wing bones embedded in the rock. The dragon was every bit as big as he'd hoped. It was a career-making find, enough to turn every other paleontologist in the world green with envy.

None was more verdant than Cope. While his rival was making splashy discoveries out West with his uncle's fortune, Cope had been stuck in New Jersey, barely scraping by. It was especially galling because Cope's father had plenty of money. He just wouldn't let Cope spend it hunting bones; he still wanted to make his son a gentleman farmer. Finally, after years of arguing, Cope forced his father to sell a tract of farmland that was supposed to be part of his inheritance. Not a moment too soon, either: shortly after the sale, Marsh published a string of papers about his new fossils, sending Cope into spasms of jealousy.

Beyond the discoveries themselves, Cope also objected to what he saw as Marsh's pernicious influence on the study of dinosaurs. Not to get too psychoanalytical, but Cope viewed dinosaurs as uncannily like himself—quick and nimble and darting. Marsh, in contrast, viewed dinosaurs as more like *him*—sluggish and methodical beasts, mostly just lumbering along. Each man of course dismissed his rival's ideas as absurd, but Marsh's sudden renown gave him more leverage to stamp his view onto paleontology. So with the proceeds from the farm sale, Cope outfit his own, rival collecting trip out West in September 1871. Achilles would now join Ajax on the field of battle.

Inevitably, word of Cope's expedition reached Marsh. Considering how Marsh had horned in on the quarries in New Jersey, it would have been the height of hypocrisy for him to protest Cope's coming west—especially given how much vaster the West was. Naturally, then, Marsh threw a fit. He spread the word among his army contacts not to trust this interloper, and when Cope arrived on the frontier, many soldiers and scouts in the forts cold-shouldered him. One fort forced him to sleep in a hay yard. Cope ignored the snubs and set off anyway.

Cope's expedition had a different feel than Marsh's. While Marsh had always loved hunting, and shot animals left and right on the trail, Cope the pacifist refused to even carry a gun, and consented to a five-soldier escort only reluctantly. While Marsh reveled in field life, thrilled to be one of the boys, Cope prudishly read out loud from the Bible each night around the campfire, ignoring the smirks, eye rolls, and disruptive belches of his crew. While both Marsh and Cope lectured from the saddle on geological formations, Cope pointed out wildflowers, too. Cope also wrote touching letters to his daughter Julia (sprinkled with Quaker "thees" and "thous"), and snagged the occasional rattlesnake to pickle in alcohol and bring back home for her.

Over the next few years, on various collecting trips, Cope's crews endured hardships galore: tornadoes; quicksand; pools of water so alkaline (or foul) that their bowels were instantly evacuated; dust storms so thick that their skin oozed grit for days afterward; swarms of insects so aggressive that they had to cake their skin with bacon grease or risk being eaten alive. As with Marsh, however, no amount of hardship could blunt Cope's excitement. He once discovered ten new fossil species in just two days, and discovered dozens of extinct species overall—weasels and mastodons, fish and gigantic tortoises. Best of all, he unearthed a pterodactyl even larger than Marsh's dragon, giving him bragging rights. To be sure, the work did take a toll on him. After thinking about ancient beasts all day, Cope would also see them in phantasmagoric dreams after sundown, an awful experience. "Every animal of which we had found traces during the day played with him at night," one companion remembered, "tossing him into the air, kicking him, trampling upon him. When I would awake him, he would thank me cordially and lie down for another attack." Yet Cope never hesitated to head out and start digging again the next morning. Such is obsession.

Overall Cope shipped literal tons of fossils back to Philadelphia. And thanks to his hare-like personality, he soon pulled ahead of his rival in terms of official discoveries. Marsh had shipped his own tons back to Yale, but even with Marsh's head start out West, the speedy Cope often beat him to the punch—pumping out journal articles and claiming priority for species that Marsh had also collected but hadn't gotten around to describing yet. Cope was also much more willing to conjure up whole new species from a jaw fragment or spare vertebra. In 1872 alone, he churned out fifty-six papers, more than one per week.

Increasingly, however, Marsh suspected that Cope wasn't relying on speed alone for an edge. Up to this point in their feud, Marsh had been the main instigator; in the classic playground defense, he started it. But the closer he looked, the more Marsh saw signs that Cope had started fighting dirty, too.

For example, there were a few cases where Marsh and Cope had unearthed the same species almost simultaneously. In poring over Cope's papers on them, however, Marsh noticed some inconsistencies with the supposed dates of discovery. Jumping to the harshest possible conclusion, he accused Cope of backdating the papers in order to secure priority for himself. (Lamely, Cope admitted the errors but blamed them on his secretary and publishers.) Around the same time, Marsh received a package from Cope with some fossils inside. In an accompanying letter Cope explained that the bones had been accidentally "abstracted" from a box of Marsh's awaiting shipment at a railroad station in Kansas. In the darndest coincidence, the missing bones had then been rerouted to Cope instead. Marsh took the letter as a taunt, and it enraged him—especially because Cope never returned the most valuable specimens in the shipment.

In response to all this, Marsh called on the powerful American Philosophical Society to censure Cope for his behavior, as well as retract old papers of his in its house journals. The society wouldn't go that far, but did agree to bar some of his future papers. Given that the society was based in Philadelphia, Cope's base of power, this boycott might have seriously damaged his career. He felt he had no choice but to neutralize the threat. When Cope's father died in 1875, he left his son $250,000 ($6 million today), and among other things,

Cope purchased a scientific journal called *The American Naturalist*. This allowed him to publish his own papers as fast as he liked, even when blocked elsewhere. As editor, Cope could also take swipes at Marsh whenever he fancied. One piece written by an ex-assistant of Marsh's called him a "scheming demagogue" and decried his "unusual elasticity of conscience." In an obituary of another assistant, Cope accused Marsh of stealing his underlings' ideas without acknowledgment. (A charge, as we'll see, with some merit.)

It didn't take long for things to escalate. Partly to cover more territory, in the mid-1870s Cope and especially Marsh began delegating more and more of the actual digging out West to platoons of professional fossil-hunters. Not surprisingly, these crews of "bone sharps" inherited their bosses' prejudices, as well as their sharp practices. Cope's men sometimes went undercover and tried to infiltrate Marsh's camps on the pretext of selling groceries. Marsh's men began spying on Cope's in turn, and they reported back to Marsh in code: Cope was "B. Jones," good luck in finding fossils was "health," and requests for money were pleas for "ammunition." Security about the location of dig sites got so tight that one of Cope's boys refused to tell even his parents where he'd be spending the summer.

Before long a few diggers defected from one camp to the other, happily selling their secrets for better pay. Others remained loyal, and would climb atop bluffs and hurl stones down at those collecting below. If anyone found a carving (like Marsh's cross) to mark a trail to revisit later, the bone sharps would efface it—then return later to grab the fossils themselves. They'd also fill old dig sites with rubble, or allegedly even dynamite them, to prevent their rivals from digging there in the future. Most egregiously, when one of Marsh's men shut down a dig site once, he crushed scores of fossil bones under his boot—pulverized them into dust—rather than risk the chance of Cope's men finding them later. The pressure got so intense that even crews on the same side got into brawls. One of Marsh's top lieutenants pulled a gun on another crew leader and challenged him to a duel.

Eventually, the war got too hot for some diggers. One resigned to herd sheep; another returned to teaching. Scientific colleagues were disgusted, too. Joseph Leidy, the discoverer of the first dinosaur in North America, actually quit dinosaur paleontology altogether, convinced that it wasn't a field for gentlemen anymore.

Still, the occasional crushed fossil notwithstanding, the field of paleontology benefitted mightily from the rivalry. Knowing they had competition, the crews pushed harder and ranged wider than they ever would have on their own, and several iconic dinosaurs were first unearthed during this period: Triceratops, Stegosaurus, Brontosaurus,* and others. Marsh and Cope also poured their personal fortunes into outfitting expeditions and preparing specimens. Marsh alone spent $30,000 ($720,000 today) to mount an especially fine Brontosaurus, the likes of which the world had never seen. Thanks to these two rivals, in fact, paleontology became the one field where American scientists clearly outpaced the rest of the world. Unlike in chemistry or physics or biology, the best research wasn't taking place in London or Paris or Berlin, but right here at home. Sinful science can have its upsides.

---

*Yes, yes, I know—save the angry letters. Technically, we should refer to the Brontosaurus (a lovely name) as the Apatosaurus (a clumsy name). But according to some scientists, Brontosaurus might be making a comeback.

The mess over this name dates back to 1877, when Marsh conjured up the Apatosaurus based on some vertebrae and bits of pelvis. Two years later, he named the Brontosaurus on equally sketchy grounds, combining the head of a sauropod from one area with the skeleton of a sauropod from another. Although a few colleagues questioned this Frankendino, Marsh's reputation kept the Brontosaurus alive until 1975, when paleontologists reevaluated several specimens and decided that Brontosaurus and Apatosaurus were the same beast. And based on the rules of scientific nomenclature, because Apatosaurus had come first, Brontosaurus was no longer valid. (The name Brontosaurus stuck around in part because museums were slow to update their displays, so the name lingered in the public consciousness.) However, according to some scientists, Brontosaurus might be a valid species after all! They've compared various old skeletons and say that Marsh's original bone fragments are distinct enough from Apatosaurus to count as a separate species. So our beloved Brontosaurus might be making a comeback. Time will tell . . .

Other fields besides paleontology benefitted, too. Charles Darwin knew that his theory of evolution would live or die by the fossil record, and Cope and especially Marsh provided vital support. One of Marsh's specialties, ancient birds with teeth, helped prove a then-controversial theory that dinosaurs had evolved into modern birds. Equally important, Marsh could trace the evolution of the horse through twenty-eight species over 60 million years, showing its transformation from a four-toed, fox-sized animal into the majestic hoofed steeds of today. When Darwin's bulldog, Thomas Henry Huxley, visited Marsh to inquire about horse evolution, he was blown away. No one had ever traced the descent of a living animal from an ancient form before. Equally impressive, every time Huxley challenged some point, or asked for proof in the form of an intermediate species, Marsh simply dispatched an assistant to fetch the very thing Huxley sought. "I believe you are a magician," Huxley finally stammered. "Whatever I want, you just conjure it up." Darwin thought Marsh no less magical, and wrote a letter praising his work on the toothed birds as "magnificent."

But if this early phase of the Cope-Marsh feud benefitted both themselves and their field, the same cannot be said of the final, Pyrrhic stages of their war.

Despite their treatment of each other, Cope and Marsh both had strong moral compasses. Cope was a pacifist who read the Bible every night. Marsh, in turn, risked his very reputation to fight the awful abuse of American Indians out West.

Marsh's crusade for Indian rights began in 1874, after a fossil-hunting trip to the Badlands of what's now South Dakota. Local tribes had refused to let Marsh enter the region at first, convinced that his "expedition" was really a cover story to steal gold from the nearby Black Hills. Tribal elders granted him passage grudgingly, and only after Marsh promised to relay their complaints about shabby treatment back to Washington. Marsh headed out in November, into land so frigid he sometimes had to chip icicles off his beard to eat dinner. To the Indians' astonishment, Marsh kept his word about not hunting for gold, returning with nothing more than wagons full of old bones.

Afterward, Chief Red Cloud took Marsh aside and showed him the basis for their complaints. In exchange for ceding land to the United States, the tribes had signed various treaties promising them foodstuffs and supplies. Red Cloud presented Marsh with that year's allotment—rancid pork, moldy flour, moth-eaten clothes, threadbare blankets. Like everyone else with a pulse, Marsh knew that the agents who distributed goods to Indians were crooked. He had no idea how crooked, though, until that moment. He was appalled, and promised to take the matter up with officials in Washington—nay, with the president himself. Red Cloud nodded and thanked Marsh, but didn't hold out much hope. He'd seen too many broken promises before. Hell, Marsh was probably in on the scam.

Defying expectations once again, Marsh traveled to Washington in 1875 for a scientific meeting and used the trip to launch a crusade against Indian agents out West. In particular, he targeted the so-called Indian Ring, a group of officials so greedy and corrupt that even the most notorious Indian-killer in the U.S. Army, George Armstrong Custer, was disgusted by them. Marsh met personally with members of the ring, and when they stonewalled his demands for reform, he called in some favors and finagled a meeting with President Ulysses S. Grant, among others. Marsh also convinced a few crusading journalists to pen exposés. Feeling cornered, the Indian Ring began spreading rumors that Marsh was a drunkard and had committed "indiscretions" out West, perhaps with the Yale boys. For once in his life, Marsh bit his tongue and didn't return fire, fearing that any mudslinging would hurt the Indian cause.

After months of work, Marsh finally blew the Indian Ring scandal open in late 1875 and forced the resignation of several prominent officials. This certainly didn't end the corruption among agents, much less the encroachment on Indian land. But Red Cloud was deeply touched by Marsh's efforts. "I thought he would do like all white men, and forget me when he went away," Red Cloud later said. "But he did not. He told the Great Father [President Grant] everything just as he promised he would, and I think he is the best white man I ever saw."

The crusade made Marsh a minor celebrity and won him the respect of many in Washington. So what did he do with his newfound prestige and moral high ground? Attack Edward Cope, naturally.

Throughout the 1870s, different agencies within the U.S. government had sponsored a series of geological surveys to produce detailed maps of the interior of the country. Both Marsh and Cope had worked on different surveys and had benefitted from the funding they provided. (Cope in fact got reprimanded repeatedly for wandering off course to hunt for fossils instead of sticking to his assigned duties.) Penny-pinchers in Congress, however, didn't like the idea of four surveys running simultaneously—it seemed redundant. In 1878 they proposed consolidating everything into one survey.

In that decision, Marsh sensed an opportunity. He'd already parlayed his celebrity in Washington into the vice-presidency of the National Academy of Sciences there; when the president of NAS died soon after, Marsh assumed control. By lucky coincidence, Congress asked NAS for advice on consolidating the geological surveys, and Marsh used every ounce of political clout he had—even meeting with President Rutherford B. Hayes—to shut down the survey that supported Cope. Marsh then got himself named chief paleontologist of the new, combined survey. In his official remarks on the consolidation, he crowed, "This is a great thing for American science." Perhaps. It was most certainly a great thing for Othniel Marsh.

Cope, meanwhile, was devastated. He'd already burned through much of his inheritance, and now his main source of outside funding had been yanked. Rashly, he took most of his remaining money and sunk it into some mining companies out West, figuring that his advanced knowledge of geology would give him an edge in selecting good investments. It didn't. Mining back then was basically legalized gambling, with the additional handicap that the casino could lie to you about the odds. Fraud and puffery abounded, and by the mid-1880s Cope was wiped out. If not for a teaching job at the University of Pennsylvania, he would have gone bankrupt.

Then came the final blow. In 1889, Cope received a letter ordering him to turn over all his fossils to the Smithsonian Institution in Washington. Cope had actually spent $75,000 of his own money collecting many of the specimens, but because he'd worked on the geological survey at the same time, the government felt it had claims to all of them. The letter came from the secretary of the interior, but Cope was convinced that Marsh was behind the whole thing.

Facing ruin, Cope decided to go nuclear. It was no secret in paleontology circles that Marsh's assistants usually ended up despising him after a few years. Marsh was stingy with pay, stole ideas, and never granted assistants any independence. Cope had already tried to take advantage of this discontent by sneaking up to New Haven for the Princeton-Yale football game one year and meeting with Marsh's men on the sly to foment rebellion. The plot failed (the assistants didn't like Cope, either), but Cope nevertheless began collecting bits of gossip in letters, which he kept in his lower-right-hand desk drawer. He called the bundle his "Marshiana," and after the threat to seize his fossils arrived, he decided to expose his rival to the world.

To do so he reached out to William Hosea Ballou, a former assistant at *American Naturalist* who now reported for the sensationalist *New York Herald.* Cope had once named a fossil after Ballou, and Ballou in turn idolized Cope: he'd once declared Cope far more brilliant than even Charles Darwin. After hearing Cope's pitch, Ballou readily agreed to write a story about how the consolidation of the geological surveys had been corrupt—and, not incidentally, to smear Marsh along the way.

Violating even the low journalistic standards of the day, Ballou interviewed several of Marsh's former assistants without telling them he was a reporter. He posed as a fellow scientist instead, who merely wanted to chitchat about a colleague. The ploy worked, and the quotes he gathered were priceless. One former assistant called Marsh's work "the most remarkable collection of errors and ignorance . . . ever displayed." Another claimed the geological survey was every bit as corrupt as Tammany Hall. Still another declared, "I never knew [Marsh] to do two consecutive honest day's work in science," then added that Marsh "has never been known to tell the truth when a falsehood would serve the purpose as well."

To give Ballou some credit, he did show the article to Marsh before publication to give him a chance to respond; he also showed it to the head of the geological survey. The survey chief penned an immediate reply, but Marsh took a different tack. He went to the University of Pennsylvania to get Cope fired from his teaching job and complete his financial ruin. When the university balked, Marsh doubled down and started digging up dirt on its president. Apparently he'd been embroiled in a sordid blackmail case, and Marsh threatened to expose him in the press if he didn't comply.

Despite Marsh's machinations, Ballou's story appeared on January 12, 1890. One historian aptly summed it up as showing "disdain for the laws of libel" and lacking even "the restraint of good taste." Virtually everyone quoted in the article denounced it—albeit without retracting what they'd said about Marsh. (Rather, they objected to Ballou's underhanded reporting methods.) Not that Ballou cared. Shamelessly, he simply gathered his sources' reactions and worked them into another hot scoop. "The fur is flying," he noted in a follow-up. "It is a very pretty fight." Nice work if you can get it.

Marsh's rebuttal—cold, precise, and nasty—appeared a week later. He insisted that it pained him to expose Cope like this, but his rival had left him no choice. Cope's charges, he claimed, were old, tired, and full of lies: "Should I attempt to give all the evidence which I have on this subject [of Cope's perfidy], verily the Sunday *Herald* with all its supplements would not contain the half thereof." (The pseudo-biblical language here was probably a dig at

Cope's Quaker speech, too.) The wickedest taunt came in answer to a charge that Marsh had plagiarized his work on horse evolution from a Russian scientist: "Kowalevsky was at last stricken with remorse and ended his unfortunate career by blowing out his own brains. Cope still lives, unrepentant." Marsh wrapped things up by reminding the world, one last time, that Cope had put the skull of a plesiosaur on its tail twenty years earlier.

In the end, the articles embarrassed both men. Rather than destroy an enemy, Cope merely created more of them, since no one trusted him now. Marsh looked petty and conniving in his replies, and soon lost his post as the geological survey's paleontologist. From a wider perspective, the scandal also sapped the fury of their rivalry. Marsh was approaching sixty then and feeling his age. Cope, nearly fifty, was in even worse shape. His wife had left him due to his financial woes, and he was sleeping alone on a cot in his house, with only a pet tortoise and his nightmare-inducing fossils for company. Then he fell ill with kidney disease, and recklessly began injecting himself with morphine, formalin (a fixative for corpses), and belladonna (a.k.a., deadly nightshade) as medicine. These home remedies did no good, and he died of renal failure in 1897. As one historian pointed out, "An inability to discharge the poisons from his system had finally killed him."

Immodestly, Cope willed his brain and skull to a colleague who studied the neurological basis of genius. According to legend, Cope's gift was also a posthumous challenge to Marsh: He was daring his nemesis to leave his own body to science in order to determine once and for all who had the bigger brain. Regardless of whether that's true, Marsh didn't take the bait. He died in 1899 and was buried in Connecticut. He had just $186 of his uncle's fortune left; he'd plowed every other dollar into his beloved fossils.

In one of the *Herald* articles, a geologist said something about Cope that applied equally well to both men. "If he could [only] be made to realize that the enemy which he sees forever haunting him as a ghost is himself."

For all their demons, though, it's hard to overestimate the impact Marsh and Cope had on natural history. At the start of the 1860s, scientists worldwide knew of roughly a dozen genera of dinosaurs. Marsh discovered nineteen genera by himself, and eighty-six species. Cope added twenty-six more genera and fifty-six species,* and wrote an incredible 1,200 papers overall. (This is supposedly a record among scientists; a list of his collected publications fills 145 pages.) And while Marsh won the species count, Cope's ideas on dinosaur biology have triumphed over his rival's. Marsh always viewed dinosaurs as the reptilian equivalent of himself—slow and plodding, an idea that reigned for a century. Nowadays, Cope's view of dinosaurs—that they were quick and nimble, like him—seems closer to the mark.

More importantly, Cope and Marsh brought about a Copernican shift in our understanding of life on Earth. Thanks to them, human beings now grasped, for the first time, just how thoroughly dinosaurs dominated our planet once, and for how long they did so—roughly 180 million years, six hundred times longer than the span of *Homo sapiens* so far. Late dinosaurs like Triceratops and Tyrannosaurus actually lived much closer in time to us today than they did to early dinosaurs like Stegosaurus, which died out 150 million years ago. This perspective also drives home the fact that, if not for a bit of luck and a big asteroid, mammals like us might still be an obscure taxon of little hairy things burrowing underground.

---

*To be sure, not all of Cope and Marsh's dinosaur species are recognized today, especially not under their original names. Perhaps more than any other science, paleontology requires the constant lumping, splitting, and reclassifying of evidence, and new species flit into and out of existence all the time. Of Cope's twenty-six genera, for instance, just three survive today. But no matter how you slice it, Cope and Marsh's taxonomic records are astounding.

The public benefitted from the rivalry as well. Marsh amassed so many crates of fossils that his successors were still unpacking them sixty years after his death, and his and Cope's collections filled museums across much of the United States. Before the Bone Wars, no one except a few academics had ever heard of dinosaurs. Cope and Marsh made dinosaurs famous—the first thing every schoolchild begs to see at a museum. They did so not just by unearthing old bones but by putting them on display and stirring people's imaginations with their writings. Consider this passage by Cope on pterodactyls, a relative of dinosaurs: "These strange creatures flapped their leathery wings over the waves and, often plunging, seized many an unsuspecting fish; or, soaring at a safe distance, viewed the sports and combats of more powerful saurians of the sea. At night-fall, we may imagine them trooping to the shore, and suspending themselves to the cliffs by the claw-bearing fingers of their winglimbs." The man could practically *see* these beasts, in his dreams and otherwise, and like a true visionary, he imparted what he saw to the rest of the world.

In the end, certain aspects of the Bone Wars have an illicit thrill to them—the double-dealing and sabotage, the defections and secret codes. Because no one really got hurt, and because science benefitted so much overall, we can chuckle about the sins of Marsh and Cope today. The same cannot be said for the next few stories, which take us into the twentieth century. The victims there weren't cranky academics, but medical patients who'd put their trust in—and were betrayed by—the very people they'd been seeking help from: physicians who'd sworn an oath to Do No Harm.

7

OATH-BREAKING: ETHICALLY IMPOSSIBLE

Smoking bans. Organic farming. Food free of dyes and preservatives. What do all these health measures have in common? They were pioneered by Nazi doctors. That's not what we normally think of when it comes to medicine under the Third Reich, of course, but the same fixation on "purity" that inspired those nostrums also inspired many of the barbaric experiments that made Nazi physicians infamous.

The Nazis were obsessed with purity, and they feared that cigarettes, processed foods, and pesticides were contaminating the bodies of German citizens. The diabolical SS even bottled and sold mineral water. The Nazis then extended this notion of purity from individual bodies to the body politic, and grew obsessed with purging society of supposed poisons, especially Jews. (As Deputy Führer Rudolph Hess once put it, "National Socialism is nothing but applied biology.") As a corollary, Nazi doctors concluded that running medical experiments on non-Aryans was not only permitted but a moral duty: the death of such "human material" would eliminate contaminants from society, and the insights gained would boost the health and well-being of the *Volk*.

Examples of ghastly experiments under the Third Reich include: shooting people with poisoned bullets; transplanting limbs without anesthesia; rubbing sawdust and glass into open wounds to study healing; and squirting caustic chemicals into people's eyes to change the color. At least 15,000 people died in such experiments (contra chapter three, Nazi anatomists never lacked for cadavers), and 400,000 more ended up crippled, scarred, or sterile. Many of those experiments would have been illegal on *animals* under Nazi law. But unlike monkeys, dogs, and horses, Jews and political prisoners enjoyed zero legal protection.

Incredibly, many of these physicians had sworn the Hippocratic oath to "Do no harm" or something equivalent—the same oath that medical students swear today. It's one of the oldest statements of medical ethics in history, dating back to the ancient Greek doctor Hippocrates, and it would seem to preclude exactly the sorts of atrocities mentioned above. So did Nazi doctors feel they'd violated this oath? Not at all. The Hippocratic oath focuses on the behavior of doctors and is largely silent about what's best for patients; it simply trusts doctors to look after them. Which is all well and good in a normal ethical landscape. But in 1930s Germany, a collectivist ethos took hold—a crude utilitarianism that ignored individual rights and promoted the "rights" of the race instead. Physicians bought into this ethos as much as anyone: As one historian noted, doctors "joined the Nazi party earlier and in greater numbers than any other professional group." As healers, they especially relished Nazi rhetoric about "curing" society's ills and eliminating "cancerous" Jews, gypsies, and homosexuals. In other words, doctors simply shifted the meaning of the Hippocratic oath from "Do no harm to patients" to "Do no harm to society," and acted accordingly. As one of their number bluntly put it, "My Hippocratic oath tells me to cut a gangrenous appendix out of the human body. The Jews are the gangrenous appendix of mankind. That's why I cut them out."

Altogether, nearly half of German doctors joined the Nazi party, and their work casts a shadow on medicine to this day. Beyond the lives cut short, there are several diseases and syndromes still named after Nazi physicians who gained their knowledge illicitly. More pressing, scientists remain split over what to do with the data from experiments on unwilling prisoners—data that's unquestionably tainted but that could nevertheless save lives today.

To be clear, most Nazi medical research deserves to be buried and forgotten. There's zero medical value in, say, sewing identical twins together, as Josef Mengele once did. Even calling such work "medicine" seems obscene.*

Not every case is so simple, though. In one series of experiments, Nazi doctors forced prisoners to guzzle seawater day after day to see how long they survived. In related work, they held people down in vats of ice water with thermometers in their rectums to study hypothermia. In still other work, they locked people inside hypobaric (low-pressure) chambers to determine the effects of extreme altitude (70,000 feet). SS chief Heinrich Himmler personally requested some of this research, and there's no question it was barbaric. In the seawater study, patients got so thirsty that they began licking the floor after it was mopped for drops of moisture. In the low-pressure study, people tore their hair out in a futile attempt to relieve the pressure imbalance inside their skulls. People in the ice baths sobbed in pain as their limbs froze inch by inch; a few begged to be shot rather than endure another minute.

But the doctors had a logical if ruthless rationale for running these experiments. Pilots were often exposed to low pressures in airplanes; sailors often got marooned on desert islands without fresh water; soldiers often suffered from exposure in the winter. Doctors

*One horrific example of Nazi deviousness involved special desks that were equipped with X-ray tubes beneath the chairs. Women of "undesirable" classes would sit at the desks while they filled out some innocuous form. All the while they were secretly being bombarded with X-rays, part of a secret plot to render them sterile.

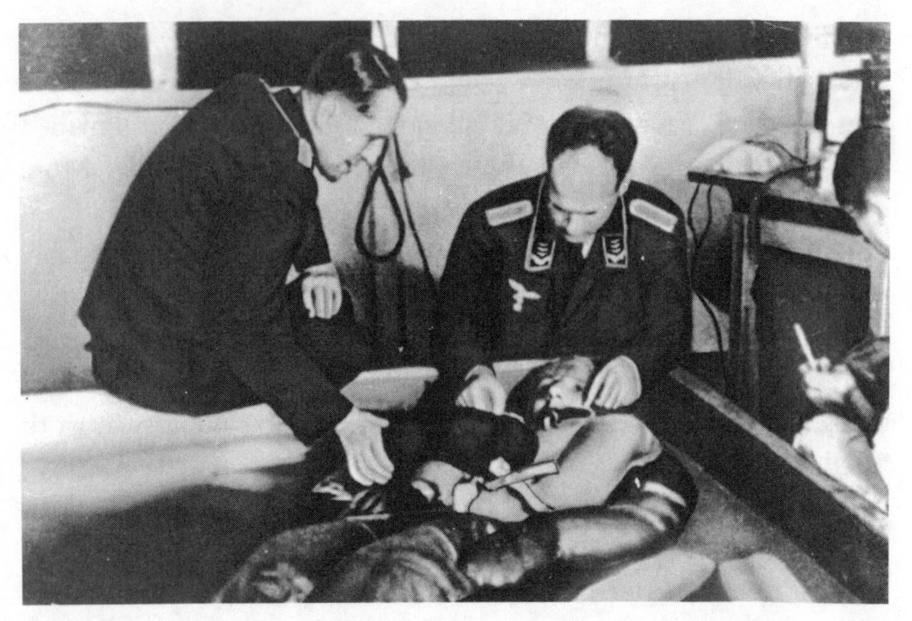

Nazi doctors hold down a prisoner in ice water during a series of barbaric experiments to study hypothermia. (Courtesy of the U.S. National Archives and Records Administration.)

wanted to know what these troops were going through physiologically, and especially how to save them. With the ice-water baths, for example, they attempted to revive people as soon as their core body temperature dropped below 80°F. Different ideas included intense sunlamps, scalding-hot drinks, heated sleeping bags, and booze. They even plucked some bewildered fellows out of the ice baths and dropped them into bed with prostitutes, who tried to get their blood flowing the old-fashioned way.

And here's the thing. For obvious reasons, no doctors since the 1940s have run similar experiments. One doctor studying hypothermia in the 1990s, for example, decided that he could not ethically lower people's core temperatures past about 93°F; anything below that was guesswork. As a result, the Nazi data is the only data we have in some cases on reviving people in extremis. Which is a problem, because the Nazi data sometimes contradicted the prevailing medical wisdom. With hypothermia, the old prevailing wisdom held that people should be warmed slowly, with their own body heat, by wrapping them in blankets or something. Doctors felt that this slow approach helped avoid shock and internal bleeding. But Nazi doctors found that passive warming like that didn't work.* Warming people rapidly and actively in hot water saved more lives.

So should modern doctors ignore that finding, due to the unethical nature of the data? Imagine a loved one—your child—falling through the ice on a river. You pluck her out, but she's barely breathing. She has blue lips, and her body temperature has dropped well below 93°F. Which revival method would you choose? The ethical but theoretical method, based on guesswork? Or the tainted Nazi method based on real data?

You could make similar arguments in other cases. In fact, some doctors maintain that saving lives today is the best way to make the victims' sacrifice mean something. And while some observers have questioned the quality of the Nazi data (it never went through peer review, for one thing), in many cases the German researchers were internationally recognized experts who knew what they were doing and set up their experiments carefully. For instance, it's common

---

*In fact, of all the rewarming methods tested, covering someone with blankets and relying on their own body heat proved the least effective. The light cradle with its sixteen intense bulbs was barely better. Vigorous rubbing of the limbs did help somewhat, but only in conjunction with hot baths. Booze turned out to be a terrible prophylactic against heat loss. It does produce a temporary feeling of warmth by sending blood rushing to the extremities, but it actually reduces the body's ability to retain heat over the long run. That said, liquor does help restore body heat once people are in a bath, since its tendency to push blood to the extremities helps relieve the burden on the heart. So if you find someone in the wild with hypothermia, by all means call a doctor first. But in the absence of that, get them into a hot tub and get them something to drink.

Incidentally, most people don't realize how shockingly close the Nazis came to covering up most of these atrocities. In fact, they might have gotten away with everything if not for the dogged efforts of one Jewish physician, Dr. Leo Alexander. For more on Dr. Alexander's incredible work, see episode 5 at samkean.com/podcast.

knowledge that you shouldn't drink seawater. German scientists therefore feared that forcing inmates to drink it would cause stress and other psychosomatic responses, which would confound their results. In response, the researchers masked the seawater's taste, making it seem far less salty. This allowed them to isolate the physiological effects of seawater alone. That's deceptive and vicious, but it's sound science.

To be sure, there are strong arguments against using Nazi data, too, both practical and moral. With the ice-water research, the inmates were often sickly and emaciated, so rewarming treatments that failed with them might still work with healthy people. Using the data could also implicitly excuse the atrocities. We should no more allow the use of improperly gathered medical results, the thinking goes, than criminal courts should allow the use of evidence seized by crooked cops.

For what it's worth, the American Medical Association, among other groups, has said that using the data might be ethical in certain circumstances—provided that there's no other way to get the information, and that anyone citing the Nazi research makes it clear that atrocities took place. Emphasizing the atrocities could even help remind us that we're not as far removed from barbarism as we like to think.

Ultimately, sixteen Nazi physicians were convicted of war crimes at the Nuremberg Doctors Trial after World War II, and seven were hanged. During the trial, American physicians and lawyers formulated ten ethical guidelines for research on human subjects, now known as the Nuremberg Code. Unlike the Hippocratic oath, the Nuremberg Code emphasizes patients' rights. Patients must give their informed consent to participate, and doctors need to take steps to minimize suffering and warn them about possible side effects and dangers. Moreover, the code states that doctors can run experiments on human beings only if there's a real medical need for the experiment and only if there's good reason to think the experiment will succeed.

In some ways, the Nuremberg Code was a tectonic shift in medical history. It made ethics an indelible part of medicine, and seventy-five years later, it still governs research on human subjects around the globe. In other ways, however, the code had little immediate impact in Allied countries. Doctors in those countries of course didn't oppose it, but to them it seemed irrelevant. Only sickos commit abuses, they thought. No doctor in a civilized nation needed such a code.

If only. In focusing on the sickos, American doctors fell into a classic psychological trap: Excusing bad behavior among their colleagues simply because it didn't sink as low as the worst possible case. *We're not as bad as the Nazis; therefore we must be okay.* In truth, rogue American doctors were running some ghastly experiments of their own during the very same months that the Nuremberg Doctors Trial was taking place. The Tuskegee and Guatemala studies lacked the sadism of the worst Nazi work. But they also proved that so-called civilized countries needed the Nuremberg Code as much as anyone.

In 1932, several white doctors working for the U.S. government's Public Health Service descended upon Tuskegee, Alabama, to study late-stage syphilis in four hundred Black men. Most of us think of syphilis as a genital disease, but if left untreated, the corkscrew-shaped bacteria that cause it can invade nearly every tissue in the body, including the heart and brain. PHS doctors wanted to study the long-term effects of this assault.

The PHS chose Tuskegee for several reasons. First, the surrounding county had an alarmingly high rate of infection, up to 40 percent in some areas. Second, the population was mostly Black, and earlier studies had suggested that syphilis affected Black people differently than white people; Black people seemed to suffer more syphilis-related heart disease, for instance, but fewer neurological complications. The

PHS venerologists wanted to determine whether those findings were true. Third, as do-good government types—many had sacrificed lucrative careers in private practice to work in public health—the PHS folks sincerely wanted to help this downtrodden community. And make no mistake, many Black people in the community welcomed their presence. This was the nadir of the Depression, and things in Tuskegee were bad. A weevil infestation had wiped out the previous year's cotton crop, and the county government had recently shuttered every public school. There was no money for health care, either, and the arrival of the PHS doctors and their promises of free physicals, free X-rays, and free bloodwork was a godsend.

To be sure, some Tuskegeans mistrusted the doctors. "White folks is sick just like us," one patient remembered thinking. So why wasn't PHS running parallel studies in white communities? Most local leaders, however, championed the research. The famed Tuskegee Institute ran medical tests to help out, and one Black doctor in town, who was also active in the civil rights movement, crowed, "The results of this study will be sought after the world over."

The study started in 1932 with physicals and bloodwork for all four hundred men. The doctors then returned every few years to follow up on them. Sometimes the follow-up involved driving the men from their farms into a clinic. Sometimes the doctors did an examination al fresco, pulling the men under the shade of a tree next to their fields and drawing some blood. The doctors then compared these four hundred men to two hundred uninfected controls, to study the burden of syphilis on their health.

One point worth noting is that the men in the study already had syphilis before the PHS arrived. Many people today believe that the men were healthy beforehand and that the doctors *gave* them syphilis via injections, which isn't true. But what the doctors did do—let the syphilis smolder untreated, sometimes for decades—was nearly as terrible.

Leaving syphilis untreated in 1932 was actually defensible. The standard treatment then involved drugs laced with arsenic

A U.S. Public Health Service doctor withdraws blood from the arm of a patient during the infamous Tuskegee syphilis study in Alabama. (Courtesy of the U.S. National Archives and Records Administration.)

and mercury. (As the saying went, syphilis was "a nighttime with Venus, a lifetime with mercury.") So heavy-metal poisoning was a real concern. Plus, killing dormant syphilis bacteria often caused them to burst open and flood the body with toxins (the so-called Jarisch-Herxheimer reaction). It was sometimes better to let sleeping syphilis lie.

However, the appearance of penicillin in the 1940s changed everything—or at least should have. Penicillin was far less toxic than earlier treatments and could cure syphilis in just eight days (arsenic and mercury cures took eighteen months). Yet even after penicillin became widely available in the 1950s, PHS doctors refused to treat the men in Tuskegee with it. Why? Because they'd set out to study the long-term effects of syphilis, and curing the men would bollix their research. So they let the germs fester instead. As one historian wrote, the study gave "the foreboding and morbid sense that the PHS watched, year after year, as the men died."

PHS doctors, of course, didn't see things that way. They considered their research virtuous. Yes, individual men in Tuskegee might be hurting, they acknowledged, but the public at large would benefit from the knowledge gained. They framed the suffering as a noble sacrifice—albeit without explaining why only Black men should make that sacrifice. Meanwhile, other PHS doctors were so focused on the biological mysteries of syphilis that they lost sight of the men as human beings. As one Tuskegee patient put it, the men were nothing but "guinea hogs" to them. One doctor even tried to suppress the publication of papers about penicillin because the drug wiped out syphilis too quickly for his liking, denying him the chance to observe the full course of the disease. ("My idea of heaven," he once said, "is unlimited syphilis and unlimited facilities to treat it.") His obsession with solving the "riddles" of syphilis blinded him to the fact that, while medical research can indeed probe some fascinating mysteries, the point of medicine isn't to satisfy intellectual curiosity. It's to heal people.

What's more, PHS doctors repeatedly lied to the men to make the study run more smoothly. Sometimes these were lies of omission: To prevent the men from getting treatment elsewhere, the doctors never told most of them that they had syphilis. (At best, the doctors made allusions to "bad blood.") There were lies of commission as well. Some men did know they had syphilis, so the doctors lured them into clinics with scams right out of a telemarketing playbook: Hurry in now, they told them in letters, or you'll miss your "last chance for special free treatment." But instead of providing treatment, they'd run dummy tests on them, or give the men excruciating spinal taps and claim to be injecting medicine.

In addition to the lies and indifference, the study also failed scientifically—which is an ethical issue in and of itself. Blood collected in the fierce Alabama heat often spoiled, and the methods used to detect syphilis were so inconsistent that doctors couldn't even say for sure whether some men had the disease. Moreover, the data analysis was inexcusably shoddy. Several controls contracted syphilis a few years into the study; similarly, some of the syphilitic men either received treatment through outside doctors or took penicillin for unrelated infections, potentially curing them. But instead of dropping these cases from the study, PHS doctors simply swapped the men back and forth between the syphilitic group and the control group—a huge no-no. In all, this scientific sloppiness rendered the study's results useless and should have torpedoed its credibility.

Now, given all the suffering in Tuskegee, harping on the data analysis might seem obtuse. But many bioethicists argue that, within medicine, sloppy science is ipso facto unethical science. It's one thing to design a terrible physics experiment and fry some vacuum pumps or whatever. No one is really harmed there. But if you're going to ask human beings to endure pain for medical research, then you're duty-bound to design the experiment properly. Otherwise the data is useless, and the pain goes for naught. Parts of the Nuremberg Code emphasize experimental design for that very reason.

All of which is to say that Tuskegee was unethical on multiple levels. The victims included more than just the untreated men, too. In most cases, late-stage syphilis isn't transmissible, but medical records made it clear that at least some of the men were contagious. By declining to tell the men they had the disease—or worse, falsely claiming to have cured it—the doctors greatly boosted the risk of the men spreading it to their wives or other sexual partners. Some of the Black scientists associated with the study suffered gravely as well. Consider Eunice Rivers.

Rivers was born in southeastern Georgia around 1900 and was no stranger to racial animus. When she was a young girl, a Black man in her hometown killed a white police officer in self-defense and fled, allegedly with her father's help. In response, white vigilantes rode up to the Rivers home on mules and shot out their windows; one bullet nearly struck and killed her. She finally escaped town in 1918 by enrolling at the Tuskegee Institute. At first she wanted to study basket-making (the school had a strong crafts program), but her father challenged her to go into science instead. She eventually became a nurse and midwife with a public-health bent, traveling from

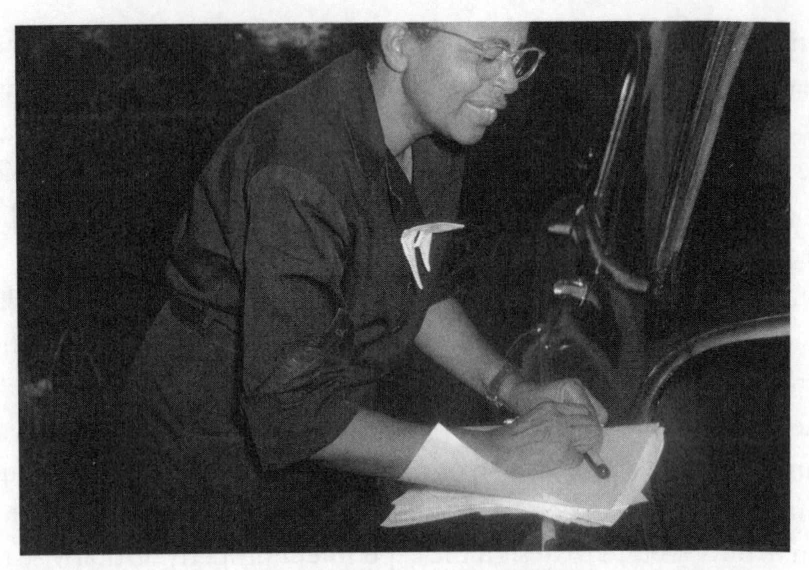

Nurse Eunice Rivers served as a key liaison for the Black community during the Tuskegee syphilis study, and was later condemned for her role in it. (Courtesy of the U.S. National Archives and Records Administration.)

house to house and giving expectant mothers hygiene tips—like spreading clean cloths or newspapers on their beds to ensure a sanitary delivery.

However meaningful the work was, Rivers longed to escape Jim Crow Alabama, and when she received an offer in 1932 to become a shift manager at a hospital in New York, she decided to take it. Then she heard about the syphilis study. The white doctors needed a liaison in the Black community, and PHS offered her a job as a scientific assistant. The chance to join a real study intrigued her, and she was so eager to make a difference in her community that she turned New York down.

Rivers played nearly every possible role in the study. She helped recruit men when it kicked off, chatting them up at churches and schools. She kept men enrolled by tracking their addresses, and she drove them to checkups in her two-door Chevy with a rumble seat. (She whooped in delight at the bawdy tales they told her on the drives; they in turn pushed her car free whenever it got mired in mudholes.) She even delivered baskets of food and clothing to them on her own time to show the men she cared. Overall, she was the one truly indispensable person on the study, and in 1958 the U.S. government rewarded her with a medal, which she was fiercely proud of. In 1953, she was also first author on a scientific paper outlining the research methods used, a rare achievement for a Black woman then.

All that said, Rivers also did some dubious things on behalf of the study. The baskets of clothing and food certainly helped the men get by, but were also an implicit bribe to keep them enrolled. Worse, one local doctor remembered her discouraging—even blocking—men in her care from getting treatment* for syphilis elsewhere, simply to preserve the integrity of the study. In some ways, then, she was no less complicit than the doctors running the experiment.

Rivers notwithstanding, there were scattered attempts to shut the study down. In 1955, a white doctor wrote a letter to PHS arguing that "[the work] cannot be justified on the basis of any accepted moral standard: pagan (Hippocratic oath), religious (Maimonides, Golden Rule), or professional (AMA code of ethics)." PHS officials ignored

*To be fair, there's anecdotal evidence that Rivers did protect at least one man enrolled in the study, a close friend of hers. She claimed he'd been lost to follow-up when in fact he lived just four blocks from the county health center. He also received a full dose of penicillin to deal with his syphilis in 1944, quite early. Furthermore, Rivers was not alone here. By 1969, the U.S. government was considering shutting the Tuskegee study down. But the Macon County Medical Society—which consisted almost entirely of Black doctors—voted to keep it going. In fact, the doctors promised that if they were supplied with a list of patients, they would withhold antibiotics from the men and refer them to Nurse Rivers instead.

Given how emotional this topic is, I'll state again that this is not—not—an attempt to shift the blame for Tuskegee onto Rivers (or the Macon County doctors). The men and women at the PHS who designed the study deserve the shame here. But Rivers did take part and was culpable in her way. I bring up her case simply because her life presented the most compelling dilemma, given how she was caught between two worlds, the Black community she lived in and the white scientific community who controlled her professional fortunes.

him. In 1969, a group of Black doctors submitted editorials to the *New York Times* and *Washington Post* denouncing the research, but the editors at both papers shrugged—it didn't seem newsworthy. Scientists didn't care, either. PHS doctors published thirteen separate papers on the study over its four decades, and made no attempt to hide what they were doing. The very first line in Rivers's paper, for instance, mentions "untreated syphilis." Indeed, that's probably the most scandalous thing about Tuskegee: that everything was out there for the public to see, and no one with any power cared.

Given that nothing was hidden, saying the study was "exposed" in 1972 isn't quite accurate. But that year, one of the whistleblowers who'd been denouncing the research—a libertarian Republican and National Rifle Association member—finally convinced an Associated Press reporter to dig into the study. When her story appeared a few months later, it exploded. Hundreds of newspapers and television stations picked it up, and the U.S Senate hauled PHS officials into hearings to grill them. The director of the Center for Disease Control (which had taken over the study from PHS) even got hanged in effigy by activists.

Most of the blame for Tuskegee fell, rightly, on the white doctors who initiated the study and refused to treat the men. But Eunice Rivers came under attack as well. When the first hostile stories appeared, she broke down weeping, and the media scrutiny got so intense that she was hospitalized for stress. Many people viewed her, as one historian put it, as either a "middle-class race traitor" or a benighted fool "who never really understands that her choosing the white doctors over the black men dooms her moral self." That bit about "choosing the white doctors" is revealing. Rivers didn't want to harm her men; many considered her a second mother, and her care packages got them through some desperate times. But as a Black woman in rural Alabama, her entire scientific career depended on the continuation of the study and her access to PHS doctors. Had she challenged them on their ethics, they almost certainly would have cut her loose.

Case studies in bioethics often take the form of melodrama. Nazi medical research is a perfect example: There are dastardly villains and innocent victims, and the outrage we feel is white-hot and uncomplicated. The PHS doctors weren't Nazi-evil, but they do fall far to the sinful side. Rivers's case is harder. She got stranded between her community and her aspirations, and she and her family struggled with what she'd done until the day of her death, and even beyond. Throughout the study, she was known as Nurse Rivers, despite having married in 1952 and becoming, officially, Eunice Verdell Rivers Laurie. But when Rivers died in 1986, her husband effectively masked her identity by putting just "Eunice V. Laurie" on her gravestone.

A half-century ago, the name Tuskegee made Black people proud. Rosa Parks was born there. George Washington Carver did his best work at the Tuskegee Institute, which Booker T. Washington founded. The Tuskegee Airmen were some of the bravest soldiers of World War II. Then PHS swept in and sullied the city's name. And again, while there's zero evidence that PHS doctors infected anyone with syphilis, that belief persists among many Black Americans today. (It's even metastasized to include other ailments: one survey in the 1990s found that more than one-third of Black people believed that the U.S. government cooked up HIV in a laboratory to commit genocide in the Black community.) Sadly but understandably, the Tuskegee study continues to undermine public health. Studies have found that, rather than submit to doctors, many in the Black community prefer to ignore warning signs for diabetes, heart disease, and other ailments until it's too late.

But the thing is, the idea of PHS doctors deliberately infecting people with STDs isn't as far-fetched as it sounds. One of the Tuskegee physicians, in fact, John Cutler, did that very thing. Just not in Alabama, but farther south, in Guatemala.

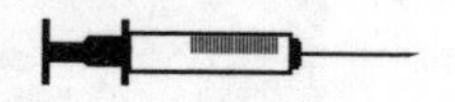

Before getting into Guatemala, it might help to examine an exact contemporary of John Cutler. This doctor was a public-health crusader. He traveled to Haiti and India during his career and worked tirelessly to improve women's access to healthcare in both countries. He arranged fellowships for gynecologists and obstetricians in developing nations to train in the United States, so they could return home and save women's lives. He also condemned the moral panic surrounding AIDS during the 1980s, and refused to demonize victims for being gay. We'll circle back to this doctor below, but he's worth keeping in mind as a moral counterpart to John Cutler in Guatemala.

After graduating from medical school in Cleveland in the early 1940s, Cutler joined PHS and began working on the surprisingly pressing issue of venereal disease in the military. VD had always run rampant among troops ("short-arm inspections" were a regular part of service), but the toll during World War II was projected to be staggering. Doctors estimated that the American military would lose seven million man-days of work to STDs every year, a labor

Dr. John Cutler, who ran the notorious Guatemala STD experiments in the 1940s on behalf of the U.S. Public Health Service. (Courtesy of the U.S. National Library of Medicine.)

loss equivalent to keeping ten full aircraft carriers at home. A few chemical prophylactics did exist then to prevent infections, but they had to be pumped down inside men's urethras, a gaggingly uncomfortable process. Many GIs skipped the treatment and just took their chances.

By 1943, however, military doctors had developed two new prophylactics—one a pill to swallow and one an ointment to spread on the outside of the penis. Cutler designed an experiment to test their effectiveness. It involved exposing 241 disease-free inmates in the Terre Haute, Indiana, penitentiary to gonorrhea (a.k.a., the clap) and seeing whether the pill or ointment prevented infection. Cutler chose Terre Haute because, as a biggish city in a region full of coal mines, there were scads of hookers around with fresh gonorrhea sores to harvest pus from.

Unlike with Tuskegee, the prisoners were fully informed of what would happen. They all signed waivers that disclosed, in plain language, the risks of exposure, and they were all promised treatment if the prophylactics failed. As for why on earth the prisoners agreed to be infected with gonorrhea, they got paid $100 each ($1,500 today), and the doctors eventually wrote letters of commendation to their parole boards. Their manhood was on the line as well. Unlike their peers on the outside, they couldn't join the military and fight Germany and Japan directly. But as Cutler slyly pointed out, they could still do their part by joining his study and keeping the troops shipshape. Nowadays, ethical codes forbid the use of prisoners in medical research because they're such a vulnerable population: they're hidden away and liable to be abused, and dangling an early release in front of them effectively coerces them and compromises their ability to consent freely. At that time, however, using prisoners was commonplace and uncontroversial. It even had some scientific advantages: the men all lived in the same environment, eliminating variation, and they were easy to keep track of and follow up on. All in all, then, Cutler designed an ethically satisfactory study by the standards of the 1940s.

If only the science had gone as smoothly. Cutler's research plan involved two steps. First, he'd smear fresh gonorrheal pus on several men's penises, in the absence of prophylactics, and measure the percentage who contracted the disease that way. This would establish a baseline rate of infection. In phase two, he'd smear the pus onto men who'd been pre-treated with prophylactics, and measure the percentage in this group who contracted gonorrhea. If the percentage in the second group was significantly lower than the baseline percentage, then the prophylactics worked.

Unfortunately, Cutler never got past the first step. He spent several months smearing gonorrhea on penises (you thought your job was bad), but the men couldn't catch the clap that way to save their lives. Without nailing down the baseline rate of infection, the study was doomed. In mid-1944, after ten futile months, PHS pulled the plug on the work, to Cutler's immense frustration.

Still, given the prevalence of VD in the military, Cutler got a second chance. By 1946 he'd transferred to a PHS office on Staten Island, where he met Dr. Juan Funes. Funes was there on a fellowship; he normally worked for the Guatemalan government's public-health office. The two got to talking, and when Funes heard about the demise of the Terre Haute study, he begged Cutler to visit Guatemala and continue the work on prisoners there. Funes did this for one reason—money. Guatemala had recently thrown off the yoke of the United Fruit Company, which had run the country as its personal colony for decades, a literal banana republic. The young nation was struggling to get on its feet, and just like in Tuskegee, money for public health was tight. Luring Cutler to Guatemala would bring U.S. doctors to train staff and U.S. dollars to buy equipment.

Cutler liked the proposal as well. One big flaw in the Terre Haute study had been the artificial method of exposure—having to smear pus on penises. Gonorrhea was normally transmitted during sex, and he reasoned that something about the sex act might allow the germs to spread more readily. Fortunately, prostitution was legal in Guatemala, even for prisoners. The women simply had to get checkups at public clinics—and it just so happened that Funes ran those clinics. He told Cutler that he could screen women there for STDs and channel them to the prison for research. Cutler could then run the same basic study as in Terre Haute, except the exposure method, sex, would be far more natural.

Despite the parallels to Terre Haute, the Guatemala research would differ in a few key ways. For one thing, penicillin had appeared on the scene by then, necessitating a change in protocol. Instead of the original prophylactic ointment, this time Cutler would mix up a paste of penicillin, beeswax, and peanut oil and smear that onto the men's penises. Funes and Cutler also decided to expand the pool of subjects to include not only prisoners but soldiers in the Guatemalan army and psychiatric patients. Similarly, they decided to look beyond just gonorrhea to syphilis and chancroid.

But the biggest difference between the two studies—and what pushed the work into the realm of sinful science—was that the doctors decided not to tell the soldiers, prisoners, and psychiatric patients that they'd be infected with STDs. Instead, they'd do it secretly. Trying to explain the science behind the study, one PHS physician insisted, would only "confuse" the poor subjects, especially the native Indians who made up the majority of prisoners. In fact, just like with Tuskegee, Cutler and company not only withheld the truth but actively lied to patients, gaining their cooperation by claiming to offer "treatments" for various diseases. The contrast is stark. When infecting American citizens, Cutler felt duty-bound to secure their consent. Guatemalans didn't warrant the same respect.

The experiments began in Guatemala City in February 1947. As planned, Funes channeled infected prostitutes to Cutler; Cutler then played pander and paired them with johns in prison. To their delight, Cutler even plied the men and women with drinks before sex. Needless to say, liquoring up research subjects wouldn't fly nowadays, but in Cutler's mind this made the sex more "natural" by simulating a meeting in the wild, namely a bar.

Still, Cutler's commitment to naturalness extended only so far. He apparently spied on the couples during sex, because he kept detailed records of how many minutes (or seconds) each man lasted—a proxy for exposure time. Then as soon as the men finished, he'd barge in and all but stick his nose in their crotches to examine the semen and vaginal fluid. There was no post-coital cuddling or cigarettes, either. For efficiency's sake, the prostitutes got less than a minute between clients. One woman had to service eight men in seventy-one minutes, with no chance to wash up in between. Most of the eventual two thousand participants in the study were adults, but one prostitute was just sixteen, and some of the soldiers were as young as ten.

Despite his initial hopes, Cutler ran into the same frustrations in Guatemala as he had in Terre Haute. Even with the booze and the wham-bam sex, the men simply weren't catching STDs at high enough rates to establish a baseline—and without a baseline, the experiments were worthless. So in his desperation, Cutler abandoned naturalistic sex and began infecting the men by hand.

It was quite the process. He first gathered some fresh venereal discharge and mixed it with nutritious beef-heart broth. Then he lured the men into his office and exposed them to this fluid in one of three ways. In the shallow exposure method, he soaked a small cotton pad in the fluid and forced it under their foreskins. (This required him, like some porno talent scout, to keep an eye out for men with meaty foreskins, which covered the pads better.) In the deep exposure method, Cutler soaked some cotton in the fluid and jammed it into the men's urethras with a toothpick. In the abrasion method, he used the tips of syringes to scratch the head of the penis until it almost bled, then slopped fluid onto the wound. Cutler also exposed uninfected female prostitutes by inserting fluid-soaked cotton balls into their vaginas and, as he reported, swishing it around "with considerable vigor." As if trying to ratchet up the creepiness, Cutler often invited his wife along to snap close-up photographs of people's genitals.

Remarkably, some of Cutler's subjects objected to these "treatments." Rather than get his penis scraped, one psychiatric patient—who it must be said, comes off as the most sane person in the room—leapt off the table and fled; it took hospital staff hours to find him. Overall, though, Cutler was quite pleased with the artificial exposure methods, which produced baseline infection rates between 50 and 98 percent.

Cutler dutifully reported all this "progress" to his superiors in Washington, who were quite impressed. One wrote to him that "your show [!] is already attracting rather wide and favorable attention up here." Another relayed a conversation he'd had with the U.S. surgeon general: "A merry twinkle came into his eye when he said, 'You know, we couldn't do such an experiment in this country.'"

Again, PHS doctors had often sacrificed lucrative careers in private practice to work in public health, and many of them came from military backgrounds to boot. In tandem, that shared background and shared sense of purpose produced a high esprit de corps within PHS ranks. Normally, a healthy esprit is a good thing. But psychologists who've studied group dynamics have found that teams with high cohesiveness and uniform backgrounds tend to make worse decisions than groups with more diversity of thought. In particular, uniform groups rarely question their own unethical behavior—or more precisely, fail to recognize they're acting unethically. As far as the homogeneous PHS was concerned, Cutler was doing a bang-up job.

Still, Cutler knew on some level that his experiments had strayed into dubious territory. Even while gloating to his superiors, he stressed the need to keep everything hush-hush. Those pleas grew more insistent after April 1947, when a short article appeared in the *New York Times*. The piece described some experiments in Baltimore and North Carolina where scientists had exposed rabbits to syphilis and immediately given them penicillin, which seemed to prevent an infection from taking hold. The reporter noted that the work held great promise in humans—but that it would be "ethically impossible" to "shoot living syphilis germs into human bodies." Meanwhile, Cutler was doing exactly that in Guatemala. Seeing his work described as "ethically impossible" gave him no pause, though. It just reinforced his suspicion that people outside PHS ranks would make trouble, so secrecy was paramount.

Tellingly, too, historians have noted that Cutler never included himself as a research subject in his experiments. That might sound like an odd criticism, but self-experimentation was quite common in medicine through the mid-1900s. Anatomist John Hunter, for instance, deliberately gave himself gonorrhea in 1767 by injecting pus into his own penis, so he could monitor the disease day by day.* However mad that sounds, Hunter at least had the courage to suffer for his science. Doctors were still doing such things in Cutler's day. In fact the Nuremberg Code carves out an exemption for dangerous research as long as there's a compelling medical need for it—and the doctors themselves serve as experimental subjects. Cutler's work was arguably compelling, but he preferred saving his own (fore)skin and exposing others.

Despite the high esprit, a few colleagues within PHS did question the Guatemala research, however tepidly. The most direct challenge involved the work with the psychiatric patients. One doctor

---

*Hunter injected his penis with venereal pus in order to determine whether gonorrhea and syphilis were the same disease or two separate ones, something that no one knew at the time. Alas, his experiment was doomed from the start, since the man he collected the pus from—unbeknown to Hunter—turned out to have both diseases. As a result, Hunter saw symptoms of each disease in himself, and erroneously concluded that syphilis and gonorrhea were in fact the same ailment. This mix-up caused all sorts of confusion until another doctor finally straightened the matter out in 1838. And while Hunter might seem heroic for experimenting on himself, he certainly didn't sidestep all the ethical issues here. For one thing, it's not clear what his then-fiancée and future wife thought of this all—or whether Hunter even told her what he was doing.

For more about self-experimentation in medicine—including some truly hair-raising tales of surgeons operating *on themselves*—see episode 20 at samkean.com/podcast.

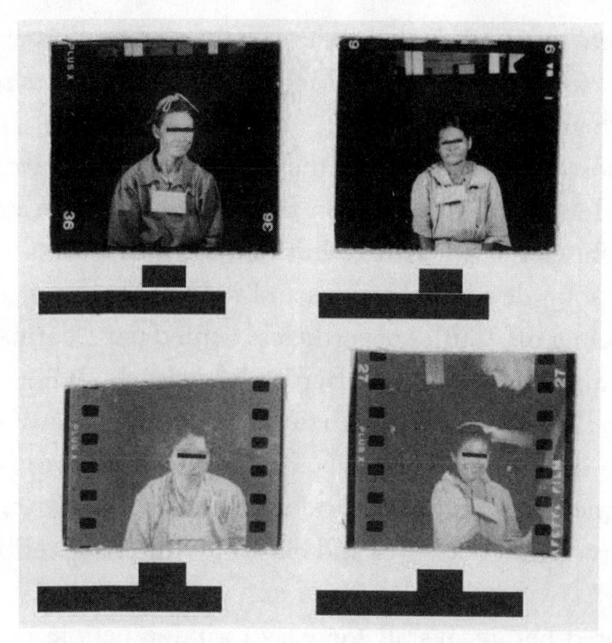

Pictures of some of the Guatemalan women deliberately infected with STDs during an experiment by the U.S. Public Health Service. (Courtesy of the U.S. National Archives and Records Administration.)

wrote to Cutler, "I am a bit, in fact more than a bit, leery of the experiment with the insane people. They cannot give consent, do not know what is going on, and if some goody organization got wind of the work, they would raise a lot of smoke." To be sure, he seems more concerned about bad press than harming people. But unlike hundreds of others within PHS, he at least raised objections and advised Cutler to stop.

The colleague was right to worry. Even considering all the other ethical lapses in Guatemala, Cutler's work in the insane asylum plumbed new lows. In exchange for some pathetically modest supplies—a projector, a refrigerator, some drugs, some plates and cups—the asylum's superintendent allowed Cutler to expose fifty psychiatric patients to STDs, including seven epileptic women who had syphilis injected into their spines. Absurdly, Cutler claimed that the women "minded the procedure so little" that they lined up "day after day" to receive the spinal injections, partly because he bribed them with cigarettes.

The most wrenching case at the asylum involved a woman named Bertha. Her age and the reason for her confinement are now lost, but in February 1948, Cutler injected syphilis germs into her left arm. She soon developed lesions and red bumps there, and her skin began peeling off. Cutler nevertheless denied her treatment for three months, and by August 23, Bertha was clearly dying. Apparently believing that he could do whatever he wanted now, Cutler proceeded to inject gonorrheal pus into her urethra, eyes, and rectum, then re-injected syphilis for good measure. Within days, Bertha was weeping pus from both eyes and bleeding from her urethra. She died August 27.

As noted before, it's all too easy to judge people in the past by the ethical standards of today and feel superior. As the saying goes, fashions in ethics change even faster than fashions in clothing, and it should give us pause to know that people in the future will probably denounce us for things we never even thought to question. But it *is* fair to judge people for violating the standards of their own day, and by that measure Cutler's "ethically impossible" work was pretty grievous. If he'd run those same experiments on Bertha in a concentration camp in Germany, he might well have been tried for war crimes.

In all PHS spent $223,000 ($2.6 million today) on Cutler's experiments before cutting off funds in 1948. Penicillin pills cured STDs so effectively that applying prophylactic peanut oil seemed pointless. A new surgeon general was taking over anyway—one presumably less prone to "merry twinkles" over ethical lapses. As a result, Cutler packed up and left Guatemala. Perhaps inevitably, given his interest in STDs, he later joined the Tuskegee study in Alabama.

Unlike the Tuskegee doctors, who blithely published their results, Cutler never wrote a word about Guatemala. That's partly because the research didn't produce any new knowledge; from a public-health standpoint, there wasn't much there. But there seems to be another, darker reason for his silence. When he left PHS in 1960, Cutler absconded with all his lab notebooks and patient charts from Guatemala, even though they were U.S. government property—a highly unusual move for a dedicated soldier like him. No one knows why he took them, but it sure smells like a cover-up, to prevent anyone else from finding out about the research. Amazingly, no one did until 2005, when historian Susan Reverby came across the notebooks at the University of Pittsburgh, where Cutler taught after leaving PHS. Had Reverby not unearthed them, and heroically gone through all ten thousand pages, they'd likely still be secret today.*

Cutler wasn't alive to see his research exposed; he died in 2003. So what sort of work had he done after Guatemala? Besides Tuskegee, he did stints in Haiti and India to help provide healthcare for women. He arranged fellowships for gynecologists and obstetricians in developing nations to train in the United States, so they could return home and save women's lives. He also condemned the moral panic surrounding AIDS in the 1980s, refusing to demonize victims for being gay.

Sound familiar? Apologies for the rhetorical ploy, but the seemingly heroic doctor at the beginning of this section—the one who championed women and minorities—was the same man who ran the Guatemala study. If the only details you knew about Cutler came from his obituary, before the Guatemala work was exposed, you'd think he was Albert Schweitzer.

So how can these two Cutlers be the same person? Perhaps he repented after leaving Guatemala, and dedicated his life to doing better. Perhaps he buried all memory of it, and refused to admit he'd done anything wrong. Perhaps he still subscribed to the crude

*I don't have room to get into the whole saga here, but the story of Susan Reverby discovering Cutler's long-suppressed work in an archive—a case that would eventually rocket all the way to the White House—is worth reading in full. Check out samkean.com/books/the-icepick-surgeon/extras/notes for the tale.

utilitarianism that as long as you were trying to help enough people overall—humankind in the abstract—you could sacrifice actual human beings along the way. (Well into the 1990s, Cutler defended the Tuskegee study on those very grounds.) Or perhaps the attempt to square Cutler One and Cutler Two misses the point. It's tempting to lump the Guatemala Cutler in with Joseph Mengele and other Nazi doctors—to dismiss him as another sicko. It's much harder to do so when we acknowledge all the good he did later. Perhaps, in the end, there is no satisfying way to reconcile the two John Cutlers.*

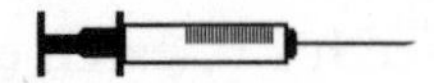

Because the Guatemala experiments were hushed up for so long, Tuskegee ended up casting a darker shadow on medical science. But as research becomes more international, sad echoes of both cases have sounded across the globe.

One controversy involved malaria vaccines. Most infectious diseases are caused by viruses or bacteria. Malaria is caused by a protozoan, a tiny but sophisticated creature with a complex life cycle. That complexity has hindered vaccine development for decades, exacerbating what's arguably the world's biggest health problem: the disease strikes 200 million people every year.

In the late 2010s, a promising new malaria vaccine appeared, and the World Health Organization (WHO) began testing it in Malawi, Ghana, and Kenya. To be sure, this vaccine, called Mosquirix, wasn't perfect. In children under seventeen months old, it cut malaria rates by only one-third. Compared to control groups, it also increased the risk of meningitis by ten times and, for mysterious

*Even Susan Reverby, who first exposed him, has resisted the urge to call John Cutler a monster. While she certainly blames him for the horrors of Guatemala, as a feminist historian she also acknowledges the great amount of good he did later in developing countries.

reasons, doubled the overall fatality rates of girls. Still, even taking those dangers into account, Mosquirix has the potential to save well over a hundred thousand lives in Africa alone every year.

The deployment of the vaccine, however, struck many critics as underhanded. Bureaucratically, WHO classified the vaccination program as a "pilot introduction" instead of a "research activity," seemingly to avoid the red tape and extra oversight that official "research" requires. Worse, officials didn't inform parents about the meningitis risk or increased fatality rates for girls. Instead, when parents showed up at clinics to vaccinate their children for other diseases, a doctor would simply ask them if they wanted a malaria vaccine as well. No one even told them that Mosquirix was experimental. WHO defended its methods by noting that parents could opt out if they chose. WHO also argued that it had gained "implied consent" by providing information to the communities beforehand about vaccines in general. But critics countered that implied consent fell well short of the "informed consent" that most research requires. As one bioethicist argued, "Implied consent is no consent at all."

As of now, this research—and the arguments over it—are still ongoing. But however iffy this all looks, the fundamental debate comes down to this: If the WHO's shortcuts speed up the introduction of the vaccine and spare a few hundred thousand lives, will the ethical sleight-of-hand have been worth it?

An even trickier case involved AIDS drugs in Uganda in the 1990s. HIV-positive women have a one-quarter chance of passing the virus to their children during pregnancy. Certain drugs can cut that rate substantially, but the drugs were too expensive for most Africans to buy—$800 per person. In addition, the treatment regimen was complicated, involving both pills and shots from healthcare workers, for both the pregnant mother and the child after birth. In response, international health officials decided to test a shorter, simpler version of the regimen in Uganda. Half the pregnant women in the study received a short course of the drug AZT, while half received an inert placebo pill as a control. Scientists then compared infection rates in the two groups to see whether the short course worked.

Placebo-controlled trials are the gold standard in medicine: scientifically, they're the best way to determine whether treatments work. But many doctors and activists were outraged at the use of placebos in Uganda. In North America, they pointed out, it's considered unethical to withhold treatment from HIV patients, even during an experimental drug trial. Instead of comparing the short course with placebos, they wanted researchers to compare the short course with the long course that people in wealthy countries would have received. Anything less, they insisted, was a double standard, and was tantamount to condemning Black babies to death.

These accusations angered the scientists running the study, including many Ugandans. The long drug course would have been too expensive for their limited budgets, they argued, greatly lowering the number of people in the study and therefore its predictive power. Furthermore, they said their opponents—mostly rich white folks in developed countries—had no idea what running studies in Africa was like, and were guilty of "ethical imperialism" for applying first-world moral standards to complex third-world situations. If not for the trial, *none* of the Ugandan women would have received any treatment. Perhaps most importantly, they reiterated that a careful scientific study with proper placebo controls was the fastest and most effective way of determining whether a treatment worked. It would therefore save the most babies in the long run.

Neither side budged, and the arguments over what to do in times of medical crisis continue today. Most recently, during the early days of COVID-19, many people wanted to unleash doctors to try all sorts of experimental drugs—despite their often-harsh side effects, side effects that could (and did) kill people who would have survived otherwise. Then again, had some of those drugs worked, we might have been spared untold amounts of heartache and pain. As mentioned earlier, many ethicists consider poorly designed medical research ipso facto unethical. But in times of crisis, even the *best* designed trials can offend people's morals. No one ever said ethics was easy.

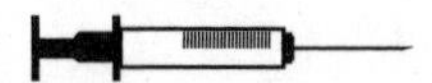

The Nazi doctors who experimented on concentration-camp inmates remain the most reviled physicians of the twentieth century, if not all time. But if they have any competition, it's an American neurologist named Walter Freeman. Unlike Josef Mengele, Freeman was no deviant, no sicko. If anything, he wanted to help people too badly—which was ultimately his undoing.

As we'll see next chapter, Freeman developed what's called the transorbital lobotomy—or, as his enemies styled it, the "icepick lobotomy." Beyond the procedure itself, it was Freeman's outsized ambition, and his forcing this "cure" on the masses, that made the lobotomy one of the most notorious medical procedures in history.

8

AMBITION: SURGERY OF THE SOUL

The story stunned Egas Moniz. It was August 1935, a discouraging year for the neurologist, and he'd arrived at the conference in London a bitter man. As soon as he heard the story of the chimpanzees, however, all his frustrations vanished.

The chimps' names were Becky and Lucy. Scientists at Yale University had been running them through some tests on memory and problem-solving. In one instance, the scientists placed a treat beneath one of two cups, then lowered a screen for a few minutes. Becky and Lucy had to remember which cup the treat was under or risk losing it. Another test involved using a short stick to drag a series of longer sticks within reach, then using the longest stick to secure another treat. Lucy aced both tasks, but Becky could never remember which cup the treat was under, and would throw epic tantrums if she missed out—hooting, slamming her fists, hurling feces, the works.

After training the chimps on these tasks, the scientists did something drastic. They surgically removed a huge chunk of the chimps' brains—their entire frontal lobes—then reran the tests to see how well Becky and Lucy coped. The results were devastating. As the Yale team reported in London, the chimps couldn't remember which cup the treat was under for more than a few seconds now, and the dragging-the-sticks task was beyond them. Losing their frontal lobes had obliterated their working memories and destroyed their ability to solve problems.

All very interesting, if a bit sad. But the insights into memory and problem-solving weren't what captivated Egas Moniz. As an aside, one Yale scientist mentioned that, after her surgery, Becky stopped throwing tantrums when she missed out on treats. She remained perfectly calm, he said, as if she'd joined a "happiness cult." All in all, removing her frontal lobes had seemingly wiped out her neurosis.

Now, this wasn't the whole story. If Becky went zen, the scientist also mentioned that Lucy went the opposite direction: After her surgery, she regressed from a calm and mature adult into a snarling, raving toddler. Removing her frontal lobes had *introduced* neurosis.

But Moniz, sitting in the audience, either missed the part about Lucy or ignored it. A vision of Becky—so calm, so serene—had seized hold of his own frontal lobes. During the Q&A afterward, he stood and asked whether brain surgery in humans might cure emotional disturbances in the same way.

The audience was shocked. Was Moniz really proposing to chop out someone's frontal lobes?

No. But he had something equally dark in mind.

Moniz might never have gotten tangled up in lobotomies if not for his illustrious family. He grew up in Portugal in the 1870s, where an uncle had filled his head with tales about his ancestors—including the original Egas Moniz, a legendary soldier who'd helped repel the invading Moors in the 1100s. Such stories inflamed the boy, and stoked a strong desire to excel and become famous himself. As a young man he attended medical school in Portugal and did a residency in neurology in Paris. Shortly afterward, at age twenty-six, he was elected to the Portuguese parliament. By middle age, he was the ambassador to Spain, and he owned a palatial estate in Lisbon with a legendary wine cellar and scads of servants, whose livery he designed himself.

Dr. Egas Moniz, the glory-hungry neurologist who invented what would become known as the lobotomy. (Painting by José Malhoa.)

To his frustration, though, his eminence in politics far outpaced his eminence in medicine. In fact, when he won an appointment in neurology at a prestigious university in Lisbon, people sneered that he'd gotten the job because of his political connections, not his scientific acumen. Such talk stung him.

Then his health began sagging. Thanks to his sumptuous lifestyle, he'd long suffered from gout in his hands. This painful joint condition turned even handshakes into excruciating ordeals, and curbed his ability to work with patients. He also gained a substantial amount of weight in middle age, leaving him looking puffy and sad.

Unable to treat patients anymore, Moniz channeled his ambitions into developing new medical procedures instead. At that time, the 1920s, doctors could examine people's bones using X-rays, but they had no good way to peer inside soft tissues. So a few scientists in France came up with what's now called the angiograph. It involved injecting opaque liquids into people's bloodstreams, liquids full of dissolved metal ions. A quick X-ray would bounce off the fluids, allowing doctors to see the contours of vessels and organs. Outside of gory accidents, this was the first-ever peek inside the guts of a living human being—a huge breakthrough.

Moniz threw himself into angiograph research, racing to secure the first pictures of the brain. He started with cadaver work. His assistant (who handled the instruments, given Moniz's hands) would take a cadaver and pump opaque fluid into its brain. Next the assistant detached the head, probably with a saw, and jumped into Moniz's chauffeured limousine with it; the limo then spirited the head across town to where the X-ray equipment awaited. For weeks, Moniz later recalled, he lived in dread of a car accident. He could practically see the severed head tumbling out onto the pavement, exposing his macabre experiments.

After the cadaver work, Moniz and his assistant graduated to live patients. But the fluids they injected (e.g., strontium bromide, sodium iodide) often leaked into the surrounding tissue, causing neurological problems like drooping eyes and seizures. One patient died. Shaken but undeterred, Moniz switched solutions and kept fiddling. In June 1927, he finally captured some gorgeous pictures of the arteries and veins that serve the brain. He even pinpointed a tumor near the pituitary gland of one patient based on the branching of vessels there.

These images were a big deal, and Moniz knew it. He worked hard to establish his priority, pumping out two dozen papers on angiographs in 1927 and 1928. Presumptuously, he also asked two colleagues to nominate him for the Nobel Prize, which they did—albeit grudgingly, apparently reluctant to refuse someone so connected.

The nominations weren't enough. Because Moniz hadn't invented angiographs, other scientists considered his work somewhat derivative, and as the 1920s passed into the 1930s, Moniz could see his share of the credit waning. There was no question that cerebral angiographs saved lives, and his colleagues now honored him as a legitimate scientist. But it was hardly enough to earn him a bust in the pantheon of his ancestors.

This was the state—sixty years old, crippled by gout, depressed over his legacy—in which Moniz arrived at the London conference in 1935. In a last-ditch effort to promote himself, he set up a booth there on angiography, but little came of the effort. Instead, Moniz spent most of his time chatting to the doctor in a neighboring booth, an ambitious young American neurologist named Walter Freeman, who also worked in brain visualization. Freeman proved a better showman than the aloof Moniz (colleagues remember Freeman at other conferences calling out like a carnival barker, rallying crowds of gawkers), but the two got along well enough, parlez-vousing in French about various aspects of their work. Pretty ho-hum.

At some point during the conference, however, Moniz attended the session on the chimpanzees Becky and Lucy—and felt the entire course of his life swerve. Few other people would have made the connection. But in the story of Becky, Moniz suddenly saw the solution to one of the most vexing problems of Western society—the shameful state of its insane asylums.

In ancient and medieval times, whenever somebody lost their wits, their families took them in and cared for them. But when industrialization fractured family life in the 1700s and 1800s, the burden for care shifted onto the government, which began herding its new wards into asylums. Every major city in the Western world had a lunatic asylum by 1900, and they were all depressingly similar: loud, filthy, overcrowded. "Patients were beaten, choked, and spat on by attendants," one historian noted. "They were put in dark, damp, padded cells and often restrained in straitjackets." (A woman at one asylum was even forced to give birth in a straitjacket, and do so in solitary confinement.) At best, asylums were warehouses for human beings. At worst, they drew comparisons to concentration camps.

Psychiatrists did try to help the insane, albeit without much success. The most common treatments involved rebooting people's brains by inducing seizures and comas with drugs or electroshocks.* Some patients did benefit from these measures (really), but only

---

*In addition to insulin coma therapy and electroshock therapy, some doctors tried Freudian talk therapy on asylum patients. But it soon became clear that propping inmates on a couch and chatting through their mommy issues was impotent in the face of real insanity, which often had roots in organic brain disorders. For this reason, Moniz and Freeman doubted that talk therapy had much to offer the truly disturbed. Freeman once quipped that any halfway-decent bartender could perform the same essential function as a psychoanalyst—listening sympathetically.

A scene from the famous Bethlem Royal Hospital in England, whose nickname—Bedlam—became synonymous with the wretched state of insane asylums. (From the *Rake's Progress* series by William Hogarth.)

some. And the less said about other "treatments"—castration, injections of horse blood, refrigerated "mummy bags"—the better.

Indeed, the most depressing thing about asylums was their air of futility. Patients moaned and wept, rocked and howled, day after day, and nothing the doctors did ever made any difference. Even calling such people *patients* seems wrong, since that implies the prospect of a cure. Really, they were inmates. Some inmates couldn't have beds, because they'd smash them and impale people with the pieces. Some couldn't have clothing because they'd tear it off or soil themselves repeatedly. In some ways, these people were worse off than animals. At least animals are placidly content. These men and women were tormented by their own minds, hour after hour, decade after decade.

Suddenly, Moniz saw a way to save them. If monkeying around inside Becky's brain had ended her outbursts, then why couldn't something similar help disturbed human beings? It was worth a shot. Except, instead of removing the frontal lobes, Moniz proposed something subtler: severing the connections between the frontal lobes and the limbic system.

In human beings, the frontal lobes allow for reflection, planning, and rational thought. The limbic system processes raw emotion. These two brain regions are connected by bundles of neurons that send signals back and forth. Moniz speculated that, in the brains of lunatics, their limbic systems had gone into overdrive, revving them up and overwhelming their frontal lobes with a barrage of signals.

Now, Moniz's theory wasn't complete bunk—disturbed emotions do overwhelm the brains of some people. But that theory was built on an outdated model of the brain as a sort of hardwired electrical switchboard, with the wires connecting different parts. To Moniz, insanity resulted from faulty wiring, full of shorts and bad connections. So by clipping those bad connections, he could return the brain to equilibrium—curing insanity with the flick of a knife.

Unfortunately, Moniz didn't seem to realize that information flows both ways in the brain. Emotions can overwhelm the frontal lobes, no question. But the frontal lobes can also send signals back to the limbic system to tamp down on raw emotions and calm us. In fact, a loss of frontal-lobe control is probably what turned Lucy, the other chimp, into such a wreck after her surgery. Without feedback from the frontal lobes, her emotions ran amok.

Again, though, Moniz ignored Lucy, focusing instead on the crisp, clean story of Becky—tormented before, placid afterward. Even more powerfully, he imagined all those millions of people suffering in asylums worldwide, and he vowed to help. Let lesser doctors waste their time with seizures and electroshocks. He would, in his word, "attack" the very roots of mental illness inside the brain, through a new discipline called psychosurgery. And if it won him glory in the process, well, he certainly wasn't opposed to that.

By the mid-1930s, Moniz was in his sixties and running out of time to establish his legacy. He therefore skipped all safety tests in animals and directed his first psychosurgery—which he called a *leucotomy*—just three months after the conference in London.*

His first leucotomy patient was a sixty-three-year-old woman who'd been in and out of psych wards for decades; she suffered from crying fits and battled hallucinations and paranoia about being poisoned. Moniz instructed a neurosurgical colleague to open two holes in the woman's skull, wafers smaller than dimes. Then they slid a syringe deep into the frontal lobes and injected a tiny shot of pure alcohol—Everclear, essentially—which destroyed the surrounding cells by dehydrating and choking them.

Even considering how much of a hurry he was in, Moniz betrayed a shocking indifference to following up with his patients, to ensure that leucotomies actually worked. In this first case, he asked the woman some silly questions a few hours after surgery ("Do you prefer milk or bouillon?"), and learned that she didn't know her age or where she was. He then sent her back to the asylum a few days later, where her crying fits resumed. He nevertheless declared her cured in early 1936, based on his impression that her paranoia and delusions were less intense. By then, he'd already moved on to other patients anyway, having injected alcohol into the brains of seven more men and women. He claimed glowing results in those cases, too, based on similarly superficial analyses.

---

*The influence of the London conference on Moniz is a controversial point. Moniz later claimed he'd been working secretly on psychosurgery for years before he ever heard of Becky, and some historians believe him. But this version seems rather self-serving, and other historians dispute the idea. For one thing, Moniz also claimed to have had conversations with colleagues about psychosurgery long before London; but when asked about this, the colleagues had no recollections of any discussions. Moniz's many writings on neurology also contain no evidence that he was working on such surgery before 1935. But again, the truth remains debated.

Privately, though, Moniz feared that the alcohol was destroying more brain cells than he wanted. He decided to switch things up and start cutting instead. This new and hopefully improved technique involved sliding a thin rod deep into the flesh of the frontal lobes. A loop of wire popped out of the rod at this point, and by rotating the loop, he could "core out" some tissue there. The cutting seemed to work on the first patient, so Moniz quickly lined up a dozen more. He declared one of these patients cured of mental illness after just eleven days, which is too short to even recover from brain surgery, much less judge its success. To him, the only real hiccup occurred when one woman groaned while the assistant was coring her—possibly because, as Moniz soon learned, the wire loop had snapped while buried inside her brain.

Already by 1936, Moniz had enough material to publish a book on leucotomies. In it, he declared that a third of his patients had been cured, a third had their symptoms significantly reduced, and a third were no worse than before. Given the futility of treating mental illness then, these would have been stunning results, if true.

True or not, people wanted to believe Moniz—they wanted the hope of a cure. That was especially the case in the United States, which had scads of squalid asylums from coast to coast. Moniz's book quickly found its way into the hands of Walter Freeman, the chatty, carnival-barking neurologist who'd set up a booth next to Moniz at the London conference. Freeman wanted to help the insane just as badly as Moniz—and would prove even more reckless in his pursuit of virtue.

Freeman called himself the Henry Ford of psychosurgery—the man who took lobotomies to the masses.

In 1936 Freeman held two posts in Washington, D.C., one at George Washington University and one at a nearby insane asylum. He liked the job at GW, where he'd gained a reputation as an electrifying teacher, someone who could pack a classroom even on Saturday mornings. With his glasses, thick eyebrows, and unfashionable mustache and goatee, he resembled Groucho Marx, and he proved no less entertaining during lectures. He could draw on the blackboard equally well with both hands, and would wow his students by sketching out two different parts of the brain simultaneously, the chalk flying. More uncomfortably, he'd scour local hospitals for interesting neurology patients and parade them in front of the students for show. For instance, one old woman with dementia had essentially deteriorated into an infant, to the point that her suckling reflex reemerged. Freeman demonstrated this by having her suck greedily on a bottle, then the bowl of his pipe. ("That's a picture they'll not soon forget," he boasted in a letter.) His students, mostly male, loved the classes so much they often brought their girlfriends along. It was better and cheaper than the movies.

In contrast to teaching, Freeman's second appointment, at the insane asylum, depressed him. Every single person there seemed miserable, from the inmates to the administrators, and the waste of human potential nauseated him. So when Egas Moniz published his book on leucotomies, detailing all the supposed cures, Freeman was ecstatic. "A vision of the future unfolded" in his mind, he recalled, something akin to a religious conversion. Freeman also had a bravado streak, and this bold new psychosurgery appealed to his sense of adventure. He quickly recruited a collaborator at GW, neurosurgeon James Watts, and got busy.

Moniz's book came out in June 1936, and already by September, Freeman and Watts had their first patients on the operating table. Now, as a neurologist, not a neurosurgeon, Freeman had no business performing operations himself. But he was too much of an alpha to just sit back and watch: once Watts opened up a skull, Freeman often took over. (To be fair, Freeman was a world-class expert on brain anatomy, his knowledge far outpacing that of Watts.) At first the duo simply copied Moniz's method of coring out brain tissue with a loop of wire. Eventually they modified the surgery, ditching the loop and cutting tissue instead with what looked like a giant butter spreader—an elongated blade with a dull edge. They slid the blade into the dime-sized holes in the skull and swished it around at different angles, to sever connections between the frontal lobes and emotional centers. Because they were using new tools and a new technique, they gave the procedure a new name, the lobotomy.

Freeman and Watts churned through a patient per week during the last four months of 1936, and their results encouraged them. Roughly half their patients calmed down enough to return home to their families, which the duo considered a huge improvement over living in an asylum. Furthermore, those who remained in the asylums were much more docile. As Freeman later put it, in a somewhat different context, "The noise level of the ward went down, 'incidents' were fewer, cooperation improved, and the ward could be brightened when curtains and flowerpots were no longer in danger of being used as weapons."

To be sure, there were setbacks. In swishing the knife around, Freeman sometimes nicked a blood vessel, and one of his early patients died of a hemorrhage. Nor did patients always improve. On Christmas Eve in 1936, one alcoholic patient staggered from his bed, pulled a hat over his surgical bandages, and wandered out the front door of the hospital. After a long search, Freeman and Watts found him ringing in the holiday at a local watering hole, so drunk he could barely walk. As a result of the misadventure, Freeman missed the birth of a son. But Freeman never let setbacks bother him. In many cases he simply scheduled another lobotomy for the patient, since he clearly hadn't carved up enough tissue the first time.

To his credit, Freeman followed up on patients far more dutifully than Moniz did, and he was intellectually honest enough (at least at first) to acknowledge the limits of lobotomies. Overall, Freeman decided that the surgery did little good for schizophrenics, alcoholics, and people with criminal perversions. (Sometimes, in fact, the perversions grew worse, since the patients lost all sense of shame after surgery. Freeman once quipped that if you gave a peeping Tom a lobotomy, instead of peering into the window, he'd barge in through the front door.) Lobotomies proved much more successful on people with severe depression and other emotional disorders, blunting their dark edges and lifting their moods. Partly for this reason, most early lobotomy patients were women, who suffered from (or were at least diagnosed with) depression and emotional disorders at higher rates than men.

Freeman was also candid about the side effects of lobotomies. None of his patients were reduced to drooling, brain-dead vegetables; that's a Hollywood stereotype. But many had to relearn basic skills like eating with utensils and using the toilet. More troubling, Freeman admitted that many patients lost their "sparkle." That is, their personalities dulled, and they lost all initiative. If someone suggested an activity to them, they'd shrug and go along, but without much enthusiasm. And if not prodded into doing something, they'd just sit around for hours, staring. Loss of frontal-lobe control also unleashed their appetites. Patients would wolf down huge meals of whatever was set in front of them, then vomit and start right up eating again. Others saw their libidos spike, and would demand sex from their spouses up to six times a day in the week after surgery. (As one writer remarked, "The knife . . . dulled Hamlet but not Romeo.") Most disturbing of all was the lack of self-awareness and social grace. One man started applauding after sermons at church, hooting and hollering as if he'd just seen a vaudeville act. Other patients stopped grooming and washing. As Freeman once said (he had a way with words), his patients exhibited "the Boy Scout virtues in reverse": a decided *lack* of cleanliness, courtesy, obedience, reverence, and so on.

Freeman's most notorious failure occurred in 1941. Joseph Kennedy, patriarch of the political clan, talked Freeman into lobotomizing his daughter Rosemary, who suffered from mood swings and angry outbursts. The lobotomy left the twenty-three-year-old Rosemary unable to speak or walk at first, and drained her of all vitality. Despite having pushed for the operation, Kennedy was furious with

Freeman. Appalled and ashamed, he locked his daughter away in an institution for the rest of her life.*

Given that even the "cured" patients suffered serious side effects, lobotomies came in for some harsh criticism. One doctor declared, "This is not an operation but a mutilation." Another said, "the psychosurgeon is indeed treading on dangerous ground when he decides that a patient without a soul is happier than a patient with a sick soul." Many doctors also raised the question of whether disturbed or insane people could truly consent to radical, experimental surgery. One of Freeman's own sons once said, "Talking about a successful lobotomy was like talking about a successful automobile accident."

Freeman didn't take such criticism lying down. He loved combat and eagerly swung back at his critics, whom he viewed as namby-pambies wringing their hands over ethics instead of actually helping

---

*Rosemary's troubles dated from her birth. Her mother's water broke unexpectedly one day in September 1918, and no doctor was available to oversee the delivery. Incredibly, a nurse on hand told Mrs. Kennedy to squeeze her legs together to hold the baby in. When Rosemary started emerging anyway, the nurse shoved her back inside. As a result, Rosemary's brain was deprived of oxygen for a few minutes, and she was never quite normal; as a child she struggled to hold a spoon properly and ride a bike.

By all accounts, Rosemary was still a vivacious young woman, and she was widely considered the prettiest Kennedy daughter. But to an ambitious family, she was an embarrassment, and they confined her to a convent as a teenager. Rosemary naturally rebelled at this, turning mouthy with the nuns and sneaking off at night—possibly, they feared, to pick up men. Given the time, a pregnant daughter would doom the family's political fortunes, so Rosemary's father Joseph started looking into lobotomies. Rosemary's sister Kathleen looked into the procedure as well and actually recommended against it, but Joseph overruled her and had Rosemary lobotomized while his wife was out of town.

John Kennedy was always disturbed by the way his family had discarded Rosemary, and he pushed through a sweeping mental-health reform bill as president. The bill's goal was to shutter massive state asylums in favor of smaller, community-based centers that could provide more intimate care. Alas, states did shutter many asylums, but they neglected to replace them with the community centers, probably to save money. The spread of psychiatric drugs only accelerated the emptying of asylums, and they've all but disappeared since.

people. He had a point, too. Even his detractors had to admit that—as strange as this sounds today—many people did benefit from lobotomies. Again, few real treatments existed for mental illness then, and lobotomies at least quieted down the most severely disturbed patients. Rather than bite anyone who came near them, or slam their heads against the wall until they passed out in a bloody heap, they could now do simple, human things like eat meals with other people or go outside to get a little sunshine. In Freeman's judgment, if the procedure "enables the patient to sleep on a bed instead of under the bed, it is worthwhile." There was no curing these people, but psychosurgery gave them something akin to normalcy. For this reason, several eminent neurologists defended Freeman, and his work garnered qualified support in publications like *The New England Journal of Medicine.*

In sum, as a treatment of last resort, the lobotomy might have had a valuable place in mid-century medicine. If only Walter Freeman had been modest enough to accept such limitations.

By the mid-1940s, Freeman was starting to have doubts about frontal lobotomies. Opening up the skull was too invasive, and worsened the effects of an already debilitating procedure. Furthermore, standard lobotomies would never put a real dent in the problem of asylums. After all, there were hundreds of thousands of mental patients in the United States, and he could operate on only one per week. Even if he taught the surgery to others, they'd still need to have an anesthesiologist and neurosurgeon on hand during the procedure. Few asylums could bear these costs, so Freeman began scouting around for a cheaper, easier surgery in 1945. He soon came across, quite literally, a new angle of attack.

Rather than drill holes through the top of the skull, Freeman read up on ways to get at the frontal lobes through the eye sockets. The orbital bones behind the eyes are relatively thin, and he realized

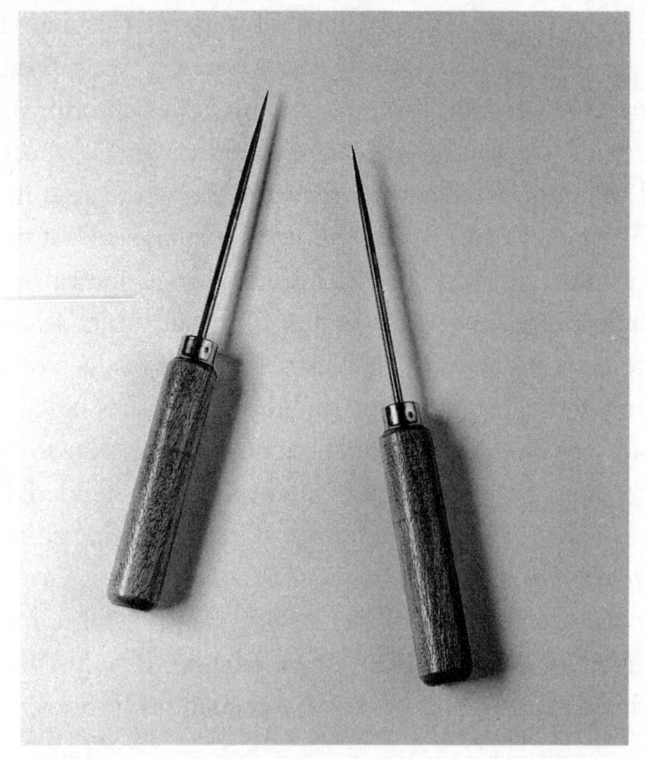

Stilettos used by Walter Freeman during his transorbital lobotomies. They were modeled after an icepick he found in his kitchen. (Courtesy of Wellcome Trust.)

that a slim rod, maybe eight inches long, could slide past the eye, puncture the orbit, and reach the brain behind. Then, by wagging the rod back and forth, he could sever the limbic system–frontal lobe connections from below. Based on the access point, Freeman called the procedure a transorbital lobotomy.

All he needed was the right tool. He got some cadavers and began experimenting with spinal-tap needles, but they proved too weak to crack the orbital bone. He finally found the perfect tool in his kitchen, when he opened a drawer one day and spotted an icepick—long, sharp, sturdy. A few cadaver tests confirmed his hunch. Suitably armed, he began scouting around for patients.

He did so secretly, however, since his partner James Watts didn't approve of the new procedure. As a surgeon, Watts was precise. He wanted to see exactly what he was cutting in the brain, not jab blindly with an icepick. Ticklishly for Freeman, he and Watts shared an office in D.C., which made operating in secret a bit awkward. Freeman nevertheless started sneaking patients upstairs to his chambers and performing transorbital lobotomies on the sly.

The procedure went like this. For "anesthesia," Freeman pulled out an electroshock machine the size of a cigar box and attached the leads to the patient's skull. A few zaps knocked her senseless. (Virtually all asylums had electroshock machines, so Freeman felt confident they could conk their patients out this way.) When the patient was insensate, Freeman pinched one of her eyelids and tented it upward, exposing the moist pink tissue beneath. Then it came time to pierce the socket. For later operations, Freeman employed a custom-made stiletto that he bragged could "practically lift a door off its hinges without it either breaking or bending." But for the first few lobotomies in his office, Freeman used his trusty kitchen icepick. He'd drop to one knee for leverage, then ease the tip of the stiletto into the tear duct. When he felt the resistance of bone behind it, he'd grab a hammer and start tapping, until he heard a *crack*. Once the tip slid into the brain, he'd swing the handle side to side at different angles to complete the lobotomy. Then it was on to the other socket. The operation rarely took longer than twenty minutes, and the patient often headed home within an hour. A few days later, two black eyes would bloom on her face—real shiners. Beyond that, if everything went smoothly, she felt minimal discomfort or pain.

Of course, things didn't always go smoothly. The electroshocks induced wild, thrashing seizures sometimes, and after a few broken limbs, Freeman had to enlist his secretaries to help hold patients down. The lobotomy had its own hazards, including infections. Freeman always pooh-poohed what he called "all that germ crap," and often operated without gloves or a face mask. On one occasion, two inches of steel actually broke off inside a patient's brain, necessitating a run to the emergency room.

(Later on, rumors swirled about even worse violations. Freeman was a serial philanderer, and while there's no firm confirmation of this, colleagues suspected that he slept with patients from time to time. Perhaps not coincidentally, he twice had to wrestle pistols away from female patients who stormed into his office. There were also rumors that Freeman would summon patients in for electroshock therapy, then secretly lobotomize them while they were groggy. How he explained their telltale black eyes was never clear.)

Given all the chaos going on upstairs, Freeman's partner James Watts soon got wise to what was happening—although the two men told different stories about how he found out. Freeman claimed he was honest and open, and invited Watts up to witness the tenth transorbital. Watts claimed he walked in on Freeman accidentally and caught him red-handed. He also claimed that Freeman, shameless as ever, just shrugged upon being busted. Then he asked Watts to hold the icepick while he snapped a few photographs.

Regardless, Watts was outraged, and demanded that Freeman stop doing experimental brain surgery in their office. However reasonable a request, Freeman bristled, and they got into a heated argument. Watts would continue to champion psychosurgery over the next decade for desperate cases, as an operation of last resort. But he refused to endorse Freeman's Jiffy Lube lobotomies, and the two had a falling-out that resulted in Watts moving out of the office.

Ever affable, Freeman didn't hold the split against Watts. In fact, it proved a blessing. Free to operate openly now, he soon got to work on his master plan: to become the Johnny Appleseed of psychosurgery, and spread lobotomies across the land.

Freeman had always loved summer road trips—hopping into a car and crisscrossing the byways of America. After perfecting the transorbital lobotomy, he decided to combine his annual excursion with work. His marriage had more or less disintegrated by that point, in part due to his workaholic habits. (He would regularly come home after dark, eat a sad dinner alone in the kitchen, then pop some barbiturates to crash asleep—only to rise at 4 a.m. the next morning to start working again.) With little keeping him at home, Freeman started bouncing from asylum to asylum in the summer of 1946 to train other doctors on doing lobotomies. Now, there's no truth to the rumor that Freeman dubbed the car he used on these trips the "lobotomobile"—but probably only because he didn't think of it.* He loved irreverent jokes, and did refer to these trips in his letters as "head-hunting expeditions."

On a typical day he'd rise at dawn at some campground and drive three or four hours to a rural mental hospital. After a tour of the grounds, Freeman might give a lecture,† followed by lunch. Then it was time for the show to start—and a show it was.

Hospitals would line up a half-dozen or so patients, and Freeman would march down the row of beds, performing one lobotomy after another. Hearkening back to his days of drawing on the chalkboard with both hands, he developed a way to do ambidextrous lobotomies, with one stiletto in each fist. Freeman claimed that these double-barreled lobotomies saved time, and they probably did, but he was also showing off for the crowd of doctors and reporters following him around. He'd even glance up midway through, grinning

* Besides the name "lobotomobile," other false rumors about Freeman include the notions that: he lost his medical license at one point; he used gold-plated ice picks for lobotomies; and he went insane later in life. None of those things are true.

† One of Freeman's favorite lecture anecdotes involved a conversation he had with a patient during brain surgery. Because the brain has no nerve endings, doctors can operate on it without the patient feeling any pain. In fact, doctors often want patients awake and talking during surgery so they can monitor them and make sure they're not cutting into anything vital. Well, one day Freeman was chatting with a patient and asked him what was going through his mind at that moment. "A knife," the man answered. Freeman found this hilarious.

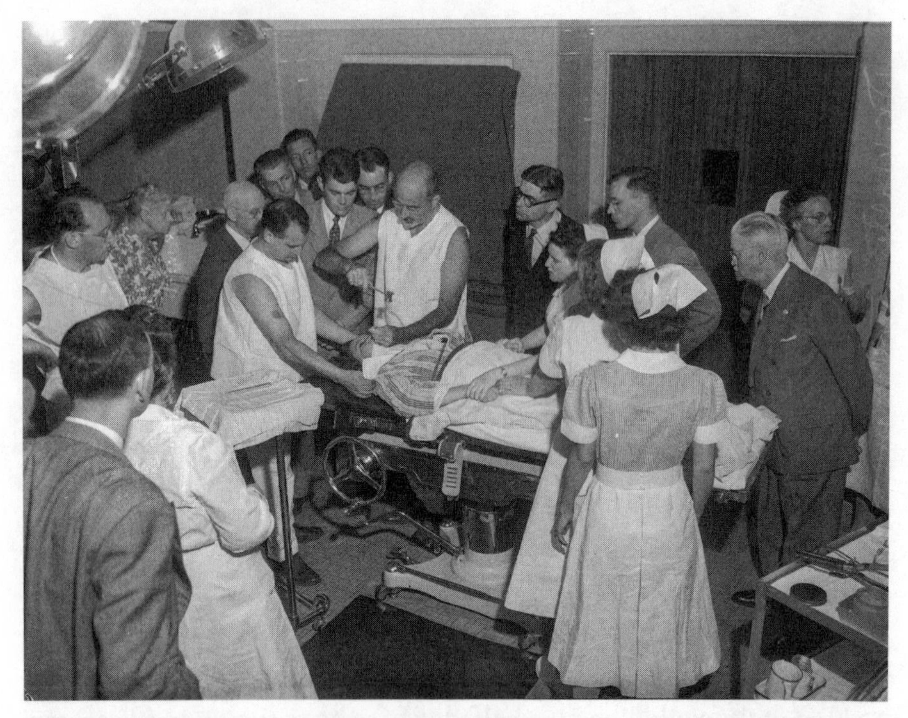

While a crowd looks on, Dr. Walter Freeman performs a transorbital "icepick" lobotomy through his patient's eye socket in 1949. Notice his bare arms, bare head, and uncovered face. (Courtesy of MOHAI, *Seattle Post-Intelligencer* Collection, 1986.5.25616.)

like Groucho, waggling the stilettos back and forth like mischievous eyebrows. One witness recalled, "I thought I was seeing a circus act . . . He was so gay, so high, so up." He also loved making people pass out. One old-time doctor—who'd worked as a medic during World War I on some of the goriest battlefields in history—fainted dead away when Freeman slammed the stiletto home through the eye socket. Freeman lectured at high schools as well, and often showed a filmstrip of a lobotomy that left half the student body light-headed. He later joked that he made more teenagers swoon than Frank Sinatra.

As always with Freeman, there were mishaps. Infections abounded, and he regularly nicked blood vessels and had to staunch hemorrhages. He also liked to photograph patients mid-surgery for documentation purposes, with the stiletto in situ. But during one operation in Iowa, gravity took over as soon as Freeman let go of the stiletto, and it plunged downward and buried itself in the patient's midbrain. The man died without regaining consciousness.

Despite the occasional fatality or three, asylum directors clamored to get Freeman through their doors. Many were no doubt sincere in wanting to help their patients, but it's hard not to be cynical about their motives when they also talked openly about all the money they'd save by sending their wards home. One calculated that, if lobotomies spread nationwide, asylums could dump 10 percent of their population and save U.S. taxpayers a million bucks a day.

The asylum tours won Freeman widespread fame—as did the fawning press coverage that accompanied them. One reporter dubbed the lobotomy the "surgery of the soul." As a result, letters from prospective patients began pouring in to his office in Washington. Most came from the miserable and wretched, who saw lobotomies as their one last shot for a normal life. Odder requests arrived, too. One man asked whether a lobotomy might cure his asthma. Another asked Freeman to lobotomize his greyhound, so it would be less skittish on the track.

Throughout the hoopla, Freeman never stopped working. He once did two dozen lobotomies in a single day; his hands would often be sore by nightfall. Then he'd say sayonara to the asylum, grab dinner somewhere on the road, and pop some barbiturates at the next campground to crash asleep. A forearm fracture one summer barely slowed him down. Even after suffering a mild stroke in 1950, he redoubled his efforts in 1951 and covered 11,000 miles that summer—and this was before the Interstate Highway System and its nice, smooth expanses. In all, he lobotomized 3,500 people over the years, bragging that he "left a string of black eyes all the way from Washington [D.C.] to Seattle."

Still, while the head-hunting expeditions made Freeman famous, they never made him rich. He paid for the travel out of pocket, charging most asylums just $20 per patient ($220 today) and often working for free. In addition, he made a determined effort to visit poor and rural areas, to help the truly neglected. This included Black communities in the South, the poorest of the poor. He had long championed—even gotten into brawls over—the rights of Black doctors to join professional medical groups. In a scary coincidence, he also tried to convince some forward-thinking doctors in Tuskegee, Alabama, to let him perform a rash of lobotomies at a veterans hospital there, since no Black neurologists lived nearby and no white neurologists would tend to the patients. The Tuskegee syphilis study was well underway by then, and if Freeman had gotten his wish, two of the twentieth century's most reviled medical practices would have collided in this unlucky town. To his disappointment, a national VA group banned lobotomies and scuttled his plan.

However charitable with the poor, Freeman could be mercenary with those who could afford it. He once charged $2,500 ($27,000 today) for a lobotomy in Chicago. In another incident, he nicked a blood vessel while operating in front of a live audience in Berkeley and watched his patient start to bleed out. "We got trouble!" he announced, and he did. Cranial pressure began to rise dangerously, and a quick neurological test (scraping a key along the bottom of the patient's foot, to test for a toe-curl reflex) showed that she was suddenly paralyzed on her right side. But instead of focusing on the patient, Freeman hopped up and left the operating theater, to shake down her husband in the waiting room. He demanded $1,000 for the trouble of fixing the mess he'd just created. One grand richer, Freeman strolled back into the operating theater, pulled what looked like a bicycle pump out of his bag, and began pumping saline into the hole in the eye socket. Moments later a clotted mass of crimson began oozing out. Freeman repeated the pumping and flushing several times, chatting all the while in his gay old way as the audience gaped. Finally, the crimson sludge oozing out got thinner, then turned pink, then clear. Freeman injected some vitamin K to encourage clotting. As a final touch, the key was dragged across the patient's foot several more times until—voilà—her toes curled. All in all, no harm done.

As disturbing as such incidents were, Freeman's real scientific sin during this period involved a shift in attitude. Neurosurgeons like James Watts restricted lobotomies to the most disturbed patients, and only as a last resort.* Freeman had approached them similarly at first. But as the years passed, and his celebrity swelled, he began promoting the operation as a prophylactic instead. That is, he started pushing for lobotomies early on during mental illness, for people who'd been institutionalized for just a few months. Even at the time, doctors knew that such people often got better on their own after a year or two; their prognosis wasn't terrible. Freeman brushed aside such numbers, arguing that psychosurgery was safer than waiting. Why not nip things in the bud, and send people home as soon as possible? He even began operating on children, some as young as four years old. The surgery of last resort had become the first line of defense.

Thanks to Freeman's tireless training of doctors, the number of lobotomies performed annually in the United States jumped tenfold between 1946 and 1949, from 500 to 5,000. Then an unexpected announcement in the fall of 1949 pushed the procedure to even greater heights.

Back in 1939, a psychotic patient had barged into the office of Egas Moniz in Portugal and shot him five times. Moniz survived, but given his gout and advanced age, he more or less retired from research, content to let Freeman and others spread psychosurgery. Still, Moniz was as indefatigable as ever in trying to secure credit for

---

*Beyond Moniz's methods (alcohol injections, cutting with a wire loop) and Freeman's methods (cutting tissue with a dull blade, the icepick penetration), several other surgeons developed their own flavors of lobotomies, including freezing brain tissue, burning it, blasting it with electricity or radiation, and aspirating it with a suction tube—which Freeman memorably described as similar to running a "vacuum cleaner over a tub of spaghetti."

himself, and in the late 1940s he once again began asking colleagues (including Freeman) to nominate him for the Nobel Prize. He received nine nominations in 1949 alone, and in the fall of that year he finally won. He thereby rose to greater prominence than perhaps any Moniz in history.

In retrospect, however, the Nobel Prize was something of a last huzzah for psychosurgery. It wasn't so much that the criticism got to be too heated. Foes did continue to attack Freeman's head-hunting expeditions, but with no better treatment options to offer as a replacement, their criticisms never gained any traction. However imperfectly, Freeman was trying to solve a real problem, and people will always flock to even a bad solution rather than sit pat. No, what finally did psychosurgery in wasn't ethics at all but $C_{17}H_{19}ClN_2S$, a compound called chlorpromazine.

Doctors in France first used chlorpromazine to treat shock, and by 1950 they were dosing inmates in asylums with it, with miraculous results. People who'd been locked away in padded cells for decades, muttering gibberish, could suddenly sit up and have conversations. Groping for an analogy, some doctors called chlorpromazine a "chemical lobotomy," but in reality the drug was far superior. It was the first true antipsychotic: a compound that didn't just deaden people (like barbiturates) but genuinely relieved their symptoms. In short, it transformed people from inmates back to patients, and few drugs in history have had a bigger social impact. Fifty million people took chlorpromazine in its first ten years on the market, and other antipsychotics like lithium quickly followed. Walter Freeman had dreamed of vacating asylums worldwide with the transorbital lobotomy; chlorpromazine actually did so. Before long, one of the most notorious features of Western society—the Bedlams that haunted every city—all but vanished.

At first Freeman praised chlorpromazine and even prescribed it to patients. Shamefully, though, as soon as the drug began to rival lobotomies, he turned against it and became a critic. To be sure, chlorpromazine wasn't perfect. As Freeman pointed out, repeatedly, the drug didn't address the root of mental illness (i.e., brain function) as much as just relieve symptoms. Indeed, many people still heard voices or hallucinated while taking the drug; the voices just didn't bother them anymore. Moreover, the drug had significant side effects: weight gain, jaundice, blurred vision, purple-tinted skin, and a Parkinson's-like shaking disorder.

Most poignant of all, chlorpromazine didn't prepare people for life after lunacy. Upon getting their minds back, patients often had no idea what year it was. The last thing one man remembered was storming an enemy trench during World War I, an event that had taken place decades earlier; then he essentially blinked, and woke up an old man. And when people left the asylum for the real world, they found that their spouses had remarried, that their skills were obsolete, that society had moved resolutely on. Even today, we're still dealing with the fallout of these drugs. Partly because of them, sanctuaries for mental illness fell out of fashion, and many people who once would have been sheltered in asylums, for better or worse, are now locked in prisons or have to fend for themselves on the streets.

On balance, however, antipsychotic drugs have done more good than harm, salvaging millions of otherwise lost lives. Beyond the social impact, these drugs have also changed our understanding of how the brain works. Back when Moniz and other scientists viewed the brain as an electrical switchboard, severing faulty "wires" with a lobotomy seemed reasonable. The advent of antipsychotics shifted our thinking. Chlorpromazine works by affecting neurotransmitters, the chemicals that send messages within the brain. As a result, scientists began to view the brain as a chemical factory, and the role of mental-health treatments was to correct chemical imbalances.

Overall, chlorpromazine had just as big an impact in psychiatry as antibiotics did in infectious-disease medicine and anesthesia did in surgery. The drug appeared just as suddenly, and it forever divided treatment into Before and After. If chlorpromazine and similar drugs had never been discovered, we'd probably still be doing lobotomies today on a limited basis—again, problems demand solutions, however imperfect. But the drugs were discovered, and for most doctors not named Walter Freeman, the choice between an imperfect drug and an icepick lobotomy was no choice at all.

Egas Moniz died serenely in 1955, confident in his legacy as both a true Moniz and a benefactor of humankind. Walter Freeman had the misfortune of being twenty years younger, and lived to see himself become a pariah.

After chlorpromazine appeared on the scene, Freeman's surgery-first attitude came to be seen as barbaric, and his brash and combative personality turned even former allies against him. In the mid-1950s he fled Washington, D.C., for Northern California, hoping for a fresh start; he even shaved off his signature beard and mustache and went barefaced for a while. But given his reputation, psychiatrists there were reluctant to make referrals to him, and he struggled to find new patients to operate on.

Instead, Freeman spent more and more time following up on old patients. He racked up gargantuan long-distance phone bills chatting with them, and tracked some as far as Australia and Venezuela—as well as the odd state prison. From these conversations he gathered reams of data, and he purchased a state-of-the-art IBM punch-card computer to sort through it all. Unlike Moniz, Freeman took follow-up seriously.

Yet however scientific this might sound, Freeman's work was too haphazard and anecdotal to have much value. For one thing, he never included control groups in his studies—asylum patients that he *didn't* operate on, and whose outcomes he could compare to lobotomy patients. Without such controls, his claims about the benefits of lobotomies were meaningless, since it's possible that his surgical patients would have improved on their own, without a lobotomy—or might even have fared better. Furthermore, given the all-too-human tendency to interpret data in the most favorable way possible, there's reason to wonder whether an ideologue like Freeman was objective in presenting his results.

Freeman performed his last lobotomy in 1967, at age seventy-two. The patient was actually one of the ten original transorbital cases he'd operated on back in his office in D.C. This was her third lobotomy, as the first two hadn't really taken. Sadly, Freeman nicked yet another blood vessel and watched her bleed out and die. His operating privileges were revoked soon after.

Instead, like Moniz, Freeman turned toward shoring up his legacy—a difficult task at that point. One reason Freeman was so dogged about following up on patients was that he could use them to deflect criticism. He never failed to mention during his talks that some of his patients had returned to productive lives as lawyers and doctors and musicians, including one at the Detroit Symphony Orchestra. When the anecdotes failed to impress, he resorted to bluster. In 1961, eleven years before his death from colon cancer, Freeman appeared onstage at a medical conference to promote lobotomies for children, and he endured some withering remarks from doctors in the audience. At which point an enraged Freeman reached down, picked up a box beside him, and spilled its contents onto a table. It contained five hundred holiday letters from grateful lobotomy patients who were still in touch with him. "How many Christmas cards did you get from *your* patients?" he demanded. It was a powerful moment—but it makes you wonder why he had the box right there. Did he suspect that this conference would be particularly hostile? Did he always carry the box around, hoping to deploy it? Or did he carry it as a balm instead, his moral shield against reproach? Regardless, the moment was pure Freeman: bold, theatrical, and defiant till the very end.

As crazy as it sounds, the Central Intelligence Agency actually commissioned a secret report on Freeman's work in the 1950s, to see whether lobotomies might help sap the zeal of communist agitators. After some thought, the agency demurred—not out of any pesky human-rights concerns, but because the surgery likely wouldn't work as intended.

As we'll see over the next two chapters, however, there were plenty of scientific abuses on both sides of the Cold War. The CIA perverted academic studies on psychological stress to develop harsher and frankly torturous interrogation techniques. The Soviet Union abused psychology, too—as well as groomed spies to ferret out secrets about the deadliest science experiment in history, the atomic bomb.

9

ESPIONAGE: THE VARIETY ACT

The pair looked like a variety act—comic opposites. One of them was thin, prim, bespectacled, and balding. He was driving a battered blue Buick that clink-clink-clinked as it rolled along. As he pulled up to the rendezvous point in Santa Fe, his partner—short, plump, and dumpy—stepped out from the shadows near a church and hopped into the passenger seat. The car took off immediately, winding to the edge of town, then up into the mountains.

It was a warm September night in 1945. After parking, the two men sat inside the car and chatted like old friends, watching the lights of the city below. Eventually the desert cooled, and they headed back to Santa Fe. Just before parting, the prim one handed his passenger a packet of papers. They shook hands warmly and, despite promises to visit, knew they'd probably never see each other again.

After the car rattled off, still clinking, the tubby man schlepped to the bus station. He had a flat-footed gait and swiveled his head as he walked, scanning the faces around him. Inside the station, he sat on a bench and tried reading a book, *Great Expectations*. But the packet of papers, which never left his grip, kept interrupting his thoughts. He also kept popping up and eyeing the crowd around him, worried he was being followed. He had good reason for being jumpy. He was a Soviet spy, and the packet contained the blueprints for an atomic bomb.

After a bus ride to Albuquerque, he caught a plane to Kansas City and went to the train station. There, he spotted an old woman and her grandson struggling to get their luggage onboard his train. Everyone else was brushing past them, so he stopped to load their bags and make sure they got settled. Unfortunately, this good deed cost him a chance to secure a seat for himself, and he had to sit on his suitcase the whole ride to Chicago.

Finally, after several delays and many, many hours, he reached New York—but did so too late to meet his Soviet contact. It was a crushing blow: their backup meeting wasn't for two weeks, which meant two more weeks of carrying the packet around, two more weeks of paranoia. But he was nothing if not disciplined. For fourteen days, he never let the papers out of his sight, even taking them grocery shopping. Indeed, there was only one place he felt safe during that fortnight—his chemistry lab.

As soon as he got his experiments running, the stress of espionage lifted a little. He could lose himself among the crucibles and test tubes and let his guard down. When he finally handed off the atomic bomb blueprints two weeks later, he buried himself in lab work again to push the matter out of his mind. Some people drink to forget troubles; Harry Gold did chemistry.

Today Gold is best known as a spy and a snitch. He accepted dozens of secret documents from a rogue physicist on the Manhattan Project—the thin, prim Klaus Fuchs—and delivered them to Soviet agents. Then, when the FBI finally caught Gold, his testimony helped put Julius and Ethel Rosenberg in the electric chair. But if anyone had asked Gold what he considered himself, the answer would have been simple. He was a chemist.

A devotion to science, and a series of misfortunes, first pushed Gold into espionage. He grew up in a rough neighborhood in South Philadelphia, where his family suffered discrimination for being Jewish. Gangs of toughs threw bricks through the windows of Jewish homes, and beat up the short, slight, bookish Gold on the way home from the library.

His father, Samson, a carpenter at a phonograph factory, had it even worse. Other workers would swipe Samson's chisels and gum up his tools with glue. His boss had a particular loathing for him, and once growled, "You son of a bitch, I'm going to make you quit." He then removed everyone but Samson from the assembly line and forced him to sand down all the wooden cabinets by himself. Samson had to work at a frantic pace, and Gold remembered him coming home with bleeding fingertips. But the young Gold admired how his old man never complained and never quit.

Samson was, however, laid off in 1931, which put Gold in a tough position. During his teenage years, Gold worked in a scientific lab at the local Pennsylvania Sugar Company; he'd started off washing spittoons and glassware before working his way up to lab assistant in just six months. He loved the work and started taking classes toward a chemistry degree. But when his father lost his job, Gold had to quit school and take a full-time position at Penn Sugar to keep them afloat. Unfortunately, the Great Depression kept getting worse, and when the company cut Gold loose in December 1932, the family faced the real possibility of losing their home.

Chemist and atomic spy Harry Gold. (Courtesy of the U.S. National Archives and Records Administration.)

After several desperate months, his friend Tom Black found him a job at a soap factory in New Jersey for $30 a week, a good wage. Gold was deliriously grateful. But there was a catch. Black was an ardent communist, and he insisted that Gold accompany him to meetings.

Although Gold leaned left politically, the communists he met disgusted him. They held tedious meetings that always ran until 4 a.m., in rooms decorated with drawings of fat plutocrats smoking cigars and sitting on piles of coins. Gold dismissed them as "despicable bohemians who prattled of free love . . . lazy bums who would never work under any economic system . . . polysyllabic windbags." A few months later, when Penn Sugar rehired Gold, he left Jersey behind and refused to attend any more meetings.

Black, though, kept hounding Gold, badgering him to join the Communist Party. Finally, to shut Black up, Gold agreed on a compromise. The Soviet Union, Black explained, needed to build up its industrial base and improve people's standard of living. The best way to do that was through science, but American firms were stingy about licensing technology abroad. Could Gold swipe some trade secrets instead?

Gold hesitated. Penn Sugar had been good to him—not many firms would have let a spittoon-washer become an assistant chemist. The company also had subsidiaries in several different chemical industries, meaning he could work in almost any field he fancied. Still, Black's pleas moved him. Gold longed to be the savior of the huddled, starving Soviet masses—people like his own family. The pitch appealed to his scientific idealism as well. By liberating trade secrets, Gold told himself, he would simply be pooling data and ideas, which is crucial to scientific progress. Soviet agents also painted science as an international brotherhood, which stood above petty things like national borders and political strife. Scientists had a duty to make the world a better place, and if American firms were too greedy to license technologies legally, then Gold had a duty to steal them.*

*One way Gold might have justified taking information from Penn Sugar was that stealing scientific trade secrets wasn't technically a crime in the 1930s. Rather, it was a civil offense. So in theory, if Penn Sugar had discovered Gold's theft, it could have sued Soviet firms in court. But good luck collecting damages.

Gold's second motivation for spying was more personal: anti-Semitism. At the time, the Soviet Union was the lone country standing up to Nazi Germany, which was of course brutally anti-Semitic. From this resistance, Gold concocted the idea that the Soviet Union was the one nation on Earth where Jewish people were truly equal—not just foes of the Nazis, but friends of Jewish people everywhere. In reality, the U.S.S.R. was just as prone to anti-Jewish prejudice as anywhere; in official KGB cables, the code word for Zionists (and according to a few sources, Jews generally) was "rats." But Gold wanted to believe otherwise. He could still remember his father's bleeding fingertips and the bricks hurled through windows, and he burned to do something "on a much wider and effective scale than . . . smashing an individual anti-Semite in the face." Supporting the great Soviet experiment was his chance to fight back.

So Gold started spying. A bit condescendingly, the Soviets code-named him "Goose" for his dumpy physique and waddling, flat-footed walk. But they quickly learned to respect his spycraft. He started off slowly, just riffling through file cabinets at work and liberating documents here and there. As the months went by, he grew bolder, nabbing papers with increasing frequency. He'd then spend hours after work, even full nights, copying the documents line by line. From various subsidiaries of Penn Sugar, he took papers on lacquers and varnishes, on solvents and detergents and alcohol. At the start, he'd never intended to take so much, but he was always thorough about his work, illicit or not. If he ever felt like quitting, he'd simply think of the poor Soviet masses and steel himself to steal more. Overall, he remembered, "I looted them pretty completely."

At first Gold just handed the pirated copies to Tom Black. Eventually he started running them up to New York himself, a task he found thrilling, since he got to meet real Soviet agents. In fact, his excitement ran so high that it energized other parts of his life as well. In addition to working full-time, he was taking evening classes at Drexel University during this period, still trying to earn that chemistry degree. His grades shot up impressively after he started spying, the B's and C's morphing into straight A's.

Still, the trips to New York got to be draining as time went on. The rendezvous often meant an overnight train ride up, not to mention hours of wandering around the city to lose potential tails. (He might sit through half a movie, for instance, before ducking out a side exit, or take the subway and dart out of his car right before the doors closed.) He'd then wait in some obscure, lonely place to meet his contact, often in snow or rain. The trips were expensive to boot—he spent more than $6,000 all told ($110,000 today)—and they decimated his health. Some weeks he barely slept, and his weight ballooned to 185 pounds.

Worst of all, he grew disillusioned with the Soviet attitude toward science. Research chemists like Gold were always searching for innovations in chemical processes, tricks and tweaks that could boost efficiency and goose production rates. Finding these breakthroughs isn't easy—hiccups and dead ends are inevitable—but most scientists accept these frustrations as the cost of progress. In contrast, the Soviets had zero patience for exploratory research or speculation; they were desperate to industrialize, and they always preferred steady, reliable processes, however old and inefficient, to even the most promising potential breakthroughs. "When I made efforts to submit material which represented work not yet in full-scale production," Gold remembered, "I had my knuckles sharply rapped." This disdain for scientific progress disappointed him, but he was submissive by nature, and his handlers bullied him into line.

Disenchanted, and increasingly weary, Gold grew to loathe his treks to New York. "It was a dreary, monotonous drudgery," he recalled, and the need to deceive his family troubled his conscience: "Every time I went on a mission . . . I must have lied to at least five or six people." (Sensing these lies, Gold's mother grew convinced her son was a libertine, stringing along girlfriends up and down the East Coast. The truth was the opposite: between work and school and spying, he had no time to date anyone, to his deep regret.) To keep sane, Gold learned to shut down his scientific mind on trips north, then pop it back on and resume the life of a chemist when he returned to Philadelphia. But each year got a little more hectic, and Gold was soon near collapse.

No matter how ragged he felt, though, Gold always found time for lab work, even if it meant a twenty-hour day. One favorite research project involved thermal diffusion, a way of using temperature differences to separate mixtures; in particular, he wanted to isolate carbon dioxide from waste exhaust to make dry ice. He described himself as a methodical chemist in the lab—less a "one-shot genius" than a plodder who made "every possible error in the book until, by the tedious process of elimination, only the correct answer remained." One afternoon he dropped a rack with twenty-two crucibles and watched a full week's worth of labor spill onto the floor. "I did not sit down and cry; nor did I go out and get drunk, as much as I wanted to," he recalled. He simply worked for two days and two nights straight to redo everything.

Gold was on the verge of quitting espionage when, in 1938, the Soviets sprung a surprise on him. He'd always longed to finish his degree, and his handler suddenly offered to pay for his tuition at Xavier University in Cincinnati. This wasn't a selfless gesture: the Soviets were developing a spy at a nearby aeronautical plant, and Gold would have to courier documents. He didn't care. He loved every second of collegiate life, putting in long hours in the lab and cheering rabidly for the Musketeer basketball and football teams. He stayed there until 1940, and later called his time in school some of the happiest years of his life.

The idyll ended on his return to Philadelphia, where Gold resumed spying. It's not clear why he did so. Perhaps he felt indebted to the Soviets for paying for his education. Or perhaps he enjoyed the companionship and sense of purpose in spying. (Indeed, the Soviets exploited his loneliness by pairing him with handlers who were also scientists. In bogus shows of camaraderie, they'd clap a hand on his shoulder and sigh, *It's too bad we have to dirty ourselves with spying. We should both be working in labs—that's what makes us happy.* As he later admitted, the communists "played me very shrewdly.") Moreover, world events compelled him to keep snooping. After Nazi Germany invaded the U.S.S.R. in 1941, the Soviets were desperate for defense help, including technical expertise. Gold hated the murderous Third Reich and resigned himself to more espionage to help the Soviet Union survive.

Thus resolved, Gold shifted from industrial espionage to military espionage. He began accepting documents and even samples of explosives from scientists in defense labs and interviewing them for reports. Managing agents like this was delicate work, requiring both technical know-how and psychological savvy. But Gold excelled. He was what the Soviets called a "disciplined athlete": a cool, reliable spy. So when the top scientific spy in the Soviet ranks—a German-born British physicist named Klaus Fuchs—transferred from England to New York in late 1943 to join the Manhattan Project, Gold was the obvious choice to handle him.

On February 5, 1944, just before 4 p.m., two men converged on a vacant lot near a playground in eastern Manhattan. One was thin and prim, wearing tweeds and glasses. He was carrying a green book and, despite the winter chill, a tennis ball.

Seeing the ball and book, a short, jowly man—wearing one pair of suede gloves and holding another in his hands—sidled up and asked for directions to Chinatown.

"Chinatown closes at 5 o'clock," the thin man answered, completing the recognition signal. And with that, physicist Klaus Fuchs and chemist Harry Gold began walking.

Gold introduced himself as "Raymond" and, after a short walk, hailed a cab. A few blocks later Gold stopped the car and hustled Fuchs into the subway to lose anyone that might be trailing them. This circuitous route eventually landed the pair at a steak house

Manhattan Project physicist and communist spy Klaus Fuchs. (Courtesy of the U.S. National Archives and Records Administration.)

on 3rd Avenue. Gold was proud of these evasive maneuvers, but Fuchs—a hardened communist from Germany, someone who'd battled Nazi thugs in the streets there—dismissed them as juvenile. He also scolded Gold for his habit of constantly swiveling his head as they walked, looking for tails. That only attracts attention, he said.

Having laid down the law, Fuchs got down to business. He explained that he was working for the Manhattan Project, which aimed to build a weapon of unprecedented power—an atomic bomb. Gold had never heard the term, and as a chemist, he grasped nuclear fission only vaguely. But one aspect of the conversation must have stirred his heart. Again, the Soviets despised most exploratory research as too speculative. Atomic fission was an exception: even though no one knew whether a nuclear bomb would work, the possibility was too important to ignore. For once, then, Gold would be handling cutting-edge research.

After this first meeting, several more followed over the next few months—in Brooklyn, Queens, the Bronx, at movie theaters, bars, museums. Every so often Fuchs handed Gold a thick envelope as they parted. Burning with curiosity, Gold would duck into a drug store and riffle through it—a huge security breach. Inside were pages filled with diagrams and mathematical derivations in a tiny, neat script. All top-secret bomb work.

Sometimes the two scientists chatted during their meetings, although each man remembered the conversations differently. Fuchs recalled professionalism—terse exchanges and strict discipline. Gold recalled a budding friendship. In between spy talk, they discussed their mutual interest in chess and classical music, and they developed a chummy cover story (what the Soviets called a "legend") to explain how they met, at a Carnegie Hall symphony. Fuchs even opened up about his sister in Cambridge, Massachusetts, who was having marital troubles. Gold in turn told Fuchs about his twin children and his wife, a redhead who once modeled for the Gimbels department store. This was a complete fabrication—the fantasy of a lonely man—but Gold let himself indulge such thoughts around his friend.

Gold also tried impressing Fuchs with his scientific knowledge. It didn't go well. During one meeting, Fuchs admitted that his team was having trouble figuring out how to enrich uranium, the first step in building a bomb core. Gold jumped in and suggested they try thermal diffusion, the process he'd been tinkering with at work. Fuchs dismissed the idea as amateurish, which stung Gold. (Little did Fuchs know that the Manhattan Project had in fact just opened a thermal diffusion plant; without it, there would have been no uranium bomb during World War II.)

In July 1944, the pair had their eighth meeting scheduled, near the Brooklyn Museum. Fuchs didn't show. This worried Gold, given how precise Fuchs always was. But they had a backup meeting scheduled a few days later near Central Park, so he took off.

Fuchs missed that meeting, too. Thoughts of muggers flashed through Gold's head, and he returned to Philadelphia distraught. He had no idea where Fuchs lived or how to contact him. An invaluable spy—and his good friend—had suddenly gone AWOL.

He needn't have worried. Fuchs was fine—better than fine, in fact. He'd just wrangled an invitation to the inner sanctum of the Manhattan Project, the weapons lab at Los Alamos.

Fuchs had left so abruptly that not even the Soviets knew where he was. Through some unknown but likely underhanded scheme, Gold's handler in New York finally tracked down the spy's last known address, near the American Museum of Natural History. He passed this information on to Gold, who bought a used copy of a Thomas Mann novel, *Joseph the Provider,* scribbled Fuchs's name and address inside, and visited the four-story brownstone on the pretext of returning it. He found the door unlocked and let himself in, but the Scandinavian couple who owned the place confronted him. Gold kept his cool and explained about the book. This assuaged the couple, but they told him that Fuchs had moved out and had left no forwarding address.

Getting desperate, the Soviets took a huge gamble and sent Gold up to Cambridge a few weeks later, to visit Fuchs's sister, Kristel. Gold arrived with a book for her and candy for her children, Fuchs's niece and nephew. He pretended to be an old friend visiting town. Alas, Kristel had no idea where her brother had disappeared to. But Gold persisted, and after several more trips up, he finally spied Fuchs sitting in Kristel's living room one afternoon in January. He must have melted with relief—seven months had passed since their last contact. The elation was short-lived, however. When Gold knocked, Kristel answered the door and promptly sent him away, telling him that Fuchs didn't care to see him just then. He should come back in two days. Gold walked away baffled and hurt—he'd taken a long, expensive train ride up to see his friend, and couldn't afford more time off to visit again.

Gold nevertheless did as he was told and returned two days later, even bringing more candy for the niece and nephew. This time Fuchs slipped outside to take a walk with Gold. On it, Fuchs apologized for sending Gold away before; Kristel's husband was home, he explained, and meeting up would have looked suspicious. The two then returned to Kristel's empty house for lunch, where Fuchs reverted to form and reprimanded Gold for compromising Kristel's security by showing up at her home so often. As a result, they'd have to rendezvous in Santa Fe from then on.

Gold groaned. That would mean even longer, more expensive trips. Wasn't there anywhere else? Fuchs said no—he was too important to get away very often. Fuchs then produced a map of Santa Fe and a bus schedule. Meet here on June 2, he said.

Just before they parted, Fuchs handed Gold a sealed envelope, which he called "extremely significant." He wasn't kidding: inside were early sketches of a plutonium bomb. Gold in turn handed Fuchs a "Christmas present" from the Soviets. It was a thin dress wallet with an envelope inside full of fives, tens, and twenties—$1,500 overall ($20,000 today). The gift disgusted Fuchs. As Gold recalled, "Fuchs held the envelope containing the $1,500 as if it were an unclean thing." I don't spy for money, he spat. This gesture delighted Gold: like him, Fuchs wasn't driven by material gain. He did convince Fuchs to take the wallet, but returned the cash to the Soviets.

Shortly before his trip to New Mexico, Gold met his Soviet handler, "John," at a bar to iron out the details. To avoid possible surveillance, John ordered Gold to take a roundabout journey to the Southwest by train and bus, with stops in California, Denver, and El Paso. For once, though, Gold stood up for himself. He had already borrowed $500 from Penn Sugar to finance the trip and couldn't afford to take more time off. He insisted on traveling directly.

If Gold won that argument, though, he would soon lose a second, more important one that day. After wrapping up the details of the Fuchs meeting, John told Gold something surprising: The Soviets had a second mole inside Los Alamos, a machinist. This mole would be on furlough in Albuquerque while Gold was visiting, so he'd be making a side trip there to pick up additional papers.

In any normal business, this would be a reasonable request. In espionage, it was anathema, a huge security risk. Gold knew this and, feeling his oats, decided to stand up for himself again: "I . . . got up on my hind legs and almost flatly refused," he remembered.

This time, John slapped him down. "I have been guiding you

idiots through every step!" he snarled. "You don't realize how important this mission to Albuquerque is."

As the tirade continued, Gold backed down and submitted, as usual. John finally gave him a sheet of onionskin paper with an address in Albuquerque on it and a last name, Greenglass. The handler then gave Gold half a Jell-O box top, which had been cut into a jigsaw shape. You'll know it's Greenglass, he said, because he'll have the matching half.

Gold's bus pulled into Santa Fe on Saturday, June 2, at 2:30 p.m. With ninety minutes to kill, he grabbed a map from a local museum and wandered along the nearby river. It looked pitiful, he thought, smaller than most creeks back home.

Fuchs arrived late in his sputtering blue Buick. They drove to a deserted road and took a short walk. Fuchs discussed his work on the new plutonium bomb but assured Gold, wrongly, that the war would end before it was ready for use against the Japanese. He then handed Gold a packet and they parted. All in all, a good, crisp meeting.

Recreation of the Jell-O boxtop recognition signal used by Harry Gold and David Greenglass. (Courtesy of the U.S. National Archives and Records Administration.)

The second meeting proved different. Gold took a bus to Albuquerque, arriving around 8 p.m., and went straight to the address on the onionskin paper, 209 High Street. He felt nervous holding documents from Fuchs and wanted to skip town as soon as possible. But David Greenglass wasn't home; he and his wife were at the movies.

Disheartened, Gold set out to find a hotel room, but got laughed at everywhere he tried. Albuquerque was surrounded by military bases, and there were never any vacancies on Saturdays. A policeman finally directed the spy to a boardinghouse, where Gold begged for a miserable cot in the hallway. Police sirens kept him awake most of the night.

The next morning Gold trudged up the staircase at 209 High Street and knocked again. When the door opened, he almost fell right back down the steps in shock. The man who answered was wearing army pants. Gold had no idea that the U.S. military had been dragged into this.

Composing himself, Gold asked if he was Greenglass. Greenglass said yes. Gold then spoke the recognition signal, "Julius sent me."

"Oh," Greenglass said, and retrieved the Jell-O box top from his wife's purse. Gold held out his half, and the pieces matched. Eager to get going, Gold asked Greenglass if he had any material ready. Greenglass said no, that he hadn't gotten around to it. Gold should come back that afternoon.

Grumbling, Gold found some breakfast and waited. When he returned, he and Greenglass took a walk in the blazing sun and made the handoff. The packet included diagrams of high-explosive lenses, one of the most crucial aspects of the plutonium bomb. Gold then handed Greenglass $500—equal to sixteen months' rent on the apartment, a small fortune to a machinist. Like Fuchs, Greenglass looked stricken at the offer—albeit for the opposite reason. Can't you do better than that?, he asked. Gold, disgusted, muttered that he'd pass the request along.

Gold caught a train out that evening, and spent the next two days rattling east, glad to have escaped. But his side trip to Albuquerque would prove costly. It just so happened that David Greenglass had a sister in New York named Ethel, who was married to a man named Julius Rosenberg.

Gold and Fuchs certainly weren't the only scientists corrupted by communism. One of Fuchs's colleagues at Los Alamos, an eighteen-year-old wunderkind named Ted Hall, spied for the Soviets as well, and a Canadian physicist actually passed along small samples of fissionable uranium. But perhaps the biggest offender was biologist Trofim Lysenko.

Lysenko was born a peasant in what's now Ukraine in 1898 and remained illiterate until age thirteen. After the Russian Revolution, he nevertheless won admission to several agricultural schools, where he began tinkering with new methods of growing peas during the long, hard Soviet winters. Although his experiments were poorly designed (and he probably faked his results), his ideas won him praise from a state-run newspaper in 1927. His hardscrabble background—people called him the "barefoot scientist"—also made him popular within the Communist Party, which glorified peasants. Officials eventually put him in charge of Soviet agriculture in the mid-1930s, vaulting him to the top of the Soviet scientific heap.

The only problem was, Lysenko had batty scientific ideas. In particular, he loathed genetics. Genetics in that era emphasized fixed traits: plants and animals had stable characteristics, encoded as genes, which they passed down to their offspring. Although a biologist in name, Lysenko condemned such ideas as reactionary—in part because the Soviet Union's greatest enemy, Nazi Germany, championed a perverted form of genetics in its promotion of a master race. But in battling right-wing fanatics, Lysenko let his own, left-wing fanaticism get the better of him, and he was every bit as unscientific as the Nazis. In fact, he went so far as to deny that genes existed. Instead, he promoted the Marxist idea that the environment alone shapes plants and animals. Put organisms in the proper setting and expose them to the right stimuli, he declared, and you can remake them to an almost infinite degree. Environment was essence.

To this end, Lysenko began programs to "educate" Soviet crops to sprout at different times of year by soaking them in freezing water, among other tricks. Crucially, he claimed that future generations of crops would remember these environmental cues and, even without being treated themselves, would inherit the beneficial traits. As science, this is nonsense—akin to cutting the tail off a cat and expecting her to give birth to tailless kittens. Nor did these tricks work on every crop. Undeterred, Lysenko was soon boasting of plans to grow lemon trees in Siberia. More importantly, he promised to boost crop production nationwide and convert the empty Russian interior into vast farms.

Such claims were exactly what Soviet leaders wanted to hear. In the late 1920s and early 1930s, Joseph Stalin had instituted a catastrophic scheme to "modernize" Soviet agriculture, forcing millions of people to join collective, state-run farms. Widespread crop failure and famine resulted. Stalin refused to change course, however, and looked in part to Lysenko's radical new ideas to remedy the disaster. For example, Lysenko forced farmers to plant seeds ridiculously close together, since according to his "law of the life of species," plants from the same "class" never competed with one another. He also forbade the use of fertilizers and agricultural pesticides.

To be clear, Stalin deserves the blame for the famines, which started before Lysenko's tenure as agricultural czar and had their ultimate roots in political factors. (Many historians even describe the famines as deliberate genocide, especially in Ukraine and Kazakhstan.) But after Stalin's crimes, Lysenko's practices prolonged the food shortages. Deaths from the famines peaked around 1932–33, but four years later, after a 163-fold increase in farmland cultivated using Lysenko's methods, food production was actually lower than before. Wheat, rye, potatoes, beets—most everything grown according to his methods died or rotted.

The Soviet Union's allies suffered under Lysenkoism, too. Communist China adopted his methods in the late 1950s and suffered through even bigger famines. Peasants were reduced to eating tree bark and bird droppings for sustenance, not to mention the occasional family member. At least 30 million died of starvation. As a corollary of Lysenko's theories—which denied the importance of genes—the Chinese government also relaxed laws against incest and consanguineous marriage. Birth defects soared as a result.

Because he enjoyed Stalin's support, Lysenko's failures did nothing to dim his star within the Soviet Union. His portrait hung in scientific institutes, and every time he gave a speech, a brass band played and a choir sang a song written in his honor.*

Outside the U.S.S.R., people sang a different tune—one of unwavering criticism. A British biologist lamented that Lysenko was "completely ignorant of the elementary principles of genetics and plant physiology . . . To talk to Lysenko was like trying to explain the differential calculus to a man who did not know his twelve times table." Lysenko in turn denounced Western scientists as bourgeois imperialists. He especially detested the American-born practice of studying fruit flies, the workhorse of modern genetics. He called such geneticists "fly lovers and people haters," apparently too ignorant to see that basic research almost always precedes practical breakthroughs.

Unable to silence his foreign critics, Lysenko tried to eliminate all dissent within the Soviet Union instead. He forced Russian scientists to renounce genetics, and those who refused found themselves at the mercy of the secret police. The lucky ones were simply dismissed from their posts and left destitute. Hundreds if not thousands of others were rounded up and dumped into prisons or psychiatric hospitals. Several were sentenced to death as enemies of the state and starved in their jail cells. Before the 1930s, the Soviet Union had arguably the best genetics community in the world. Lysenko gutted it, and by some accounts set Russian biology back a half-century.

Lysenko's grip on power began to weaken after Stalin died in 1953. By 1964, he'd been deposed as the dictator of Soviet biology, and he died in 1976. His portrait continued to hang in some institutes

*A sample lyric: "Merrily play one, accordion, / With my girlfriend let me sing / Of the eternal glory of Academician Lysenko." Perhaps it sounds better in Russian.

through the Gorbachev years, but by the 1990s, the Russian people had finally put the horror and shame of Lysenkoism behind them. Or at least they thought they had.

In 2017, four scientists in Russia wrote a journal article sounding the alarm about a resurgence in Lysenkoism. Several books and papers praising his legacy had recently appeared, bolstered by what they called "a quirky coalition of Russian right-wingers, Stalinists, a few qualified scientists, and even the Orthodox Church."

There were several reasons for the revival. For one, the hot new field of epigenetics had made Lysenko-like ideas fashionable again. But the real explanation was opposition to Western values. The four Russian scientists explained that Lysenko's latter-day disciples "accuse the science of genetics of serving the interests of American imperialism and acting against the interests of Russia." Science, after all, is a major component of Western culture. Lysenko the barefoot peasant had opposed Western science—ergo he must be a Russian hero. Indeed, nostalgia for the Soviet era and its anti-Western strongmen is common in Russia today. (See Vladimir Putin.) A 2017 poll found that 47 percent of Russians approved of Joseph Stalin's character and "managerial skills," and several of his lackeys, including Lysenko, are riding Big Joe's coattails back to popularity.

On the one hand, this rehabilitation is shocking. Genetics almost certainly won't be banned in Russia again, and support for Lysenko remains a fringe movement overall. But fringe ideas can have dangerous consequences, and the new Lysenkoism distorts Russian history and glosses over the incredible damage he did in silencing and killing colleagues—to say nothing of the thousands of farmers whose crops failed because of his doctrines. The fact that even some scientists in Russia are lionizing Lysenko shows just how deep anti-Western sentiment runs there.

That said, there is something depressingly familiar about the Lysenko revival. Even in the Western world, ideology perverts people's scientific beliefs all the time. Nearly 40 percent of Americans believe that God created human beings in their present form, sans evolution; nearly 60 percent of Republicans attribute global temperature changes to nonhuman causes. And while there's no real moral comparison between them, it's hard not to hear echoes of Lysenko in Sarah Palin's mocking of fruit-fly research in 2008. Lest liberals get too smug, several left-wing causes—genetically modified food hysteria, the "blank slate" theory of human nature—sound an awful lot like Lysenkoism redux.

To their credit, Lysenko's contemporaries Harry Gold and Klaus Fuchs eventually sobered up and realized what monsters Stalin and his scientific henchmen were—a threat to not only science but humanity. That epiphany came awfully late, though. As Fuchs once said, "Some people grow up at fifteen, some at thirty-eight. It's much more painful at thirty-eight." In the meantime, Fuchs and Gold continued to loot the Manhattan Project, and do everything they could to hand Joseph Stalin an atomic bomb.

Fuchs and Gold rendezvoused in Santa Fe again in September 1945, in the meeting described at the beginning of the chapter. World War II was over by then, but the Soviets were already ramping up for the Cold War and were even more desperate than before for atomic data.

To Gold's surprise, Fuchs revealed that night that he and the other British scientists at Los Alamos would be returning to England soon. (Fuchs's excuse for slipping up to Santa Fe was in fact to pick up booze for a going-away party for the Brits; hence the clinking in the Buick: The trunk was full of bottles.) According to Gold, they talked vaguely of him visiting Fuchs in England someday, since he'd always loved Wordsworth and Shakespeare and wanted to see their homeland. Fuchs agreed that it all sounded grand. He then handed Gold a packet with data on the Hiroshima and Nagasaki bombs.

This was also the trip where Gold missed his contact in New York, forcing him to spend two weeks carrying the papers around. The fortnight exhausted him, and given the expense and added stress of trips to New Mexico, he decided once more that he was through with espionage.

Espionage, however, wasn't through with him. In 1946, Penn Sugar laid Gold off again. He applied to the KGB for funds to open a thermal diffusion lab, and when the agency turned him down, Gold moved to New York to seek work with a fellow chemist—and fellow communist spy—named Abe Brothman. It was a huge mistake. The FBI had its eye on Brothman, and Gold's handlers had told him to steer clear of the man. But Gold either forgot their warning or ignored it, and accepted a job anyway. (When the Soviet handler John found out that Gold was working for Brothman, he screamed at Gold, in public, "You fool! You've spoiled eleven years of preparation!") Sure enough, Brothman soon ran afoul of the FBI and implicated Gold in espionage. The two were summoned to testify before a grand jury in July 1947.

Brothman was every bit as weary of Soviet demands as Gold was, and he began threatening in private to confess his role in the Soviet spy machine and bring everyone down with him. But he pulled himself together in the witness box and denied everything. Nine days later came Gold's turn to testify. The night before, Gold swung by Brothman's apartment, and they went for a drive. Gold wanted to discuss his testimony, to make sure it didn't contradict Brothman's, but every time he brought the topic up, Brothman started ranting about the death of capitalism. Gold finally gave up, and the two stopped to eat some watermelon at 4 a.m.

Gold needn't have worried: he proved every bit as deft at lying under oath as Brothman had. On the stand, he made himself out as a bumbling, absentminded chemist, someone too naïve to even know what politics was. While the FBI didn't believe either man, it couldn't poke any real holes in their stories, and both of them walked.

Still, thanks to Brothman, the FBI now had a file on Gold. Even worse, agents in England were about to reel in a spy who had much more incriminating material on him—none other than Klaus Fuchs.

Brothman paid Gold erratically, if at all. ("When there was no money, I was a partner," Gold said of his time there. "When there was money, I became an employee.") Gold also missed his family while in New York, especially after his mother died of a brain hemorrhage in September 1947. So in mid-1948 Gold quit Brothman's firm, moved back home, and took a job in the Heart Station at Philadelphia General Hospital. Not only was he doing good, solid chemistry—studying electrolyte levels in the blood and how potassium affected muscle function—he was helping save people's lives. He even met the love of his life at the hospital, a biochemist named Mary Lanning. "I had never been happier," he later said.

Over the next year and a half, Gold proposed to Lanning twice. She twice said no—but not because she didn't love him. Rather, she sensed that he was hiding something about his past, something big. For example, when he accidentally mentioned visiting Santa Fe once, he covered his tracks by saying Penn Sugar had sent him down to check out a Coca-Cola plant there—obvious baloney. No dummy, Lanning knew he was holding something back, but Gold simply couldn't reveal the truth and risk her thinking less of him. The couple finally broke things off, because he feared that if they married and he was exposed, it would ruin her life.

He was right to worry. In September 1949, four years after his last meeting with Fuchs, Gold answered the door at his home one Saturday night to find a man with a thick accent standing there. Not knowing who he was, Gold moved to shut the door. Before he could, the man—a Soviet agent—recited a code phrase, which stopped Gold short. He'd been hearing rumors about uncooperative spies being tracked down and murdered, and no matter how much he wanted to blow this man off, he felt it safest—for him and his family—to play along. After some discussion in the kitchen, Gold agreed to visit the agent in New York two weeks later. They rendezvoused during a downpour. To Gold's shock, the agent demanded that Gold defect to Eastern Europe. He refused to explain why.

Everything became clear a few months later. On February 2, 1950, Klaus Fuchs was arrested in England and confessed. The United States was still reeling from the news that the Soviet Union had detonated a nuclear bomb the previous August, and the capture of an atomic spy made headlines worldwide. Seven days later, during a speech in Wheeling, West Virginia, Wisconsin Senator Joseph McCarthy waved around a piece of paper that, he claimed, listed 205 communists who'd infiltrated the State Department. He'd been itching for months to begin an anti-communist purge, and Fuchs had given him a golden opportunity. Fuchs's arrest also convinced Harry Truman—who'd previously announced plans only to study the feasibility of a hydrogen bomb—to fully commit to building one.

One part of Fuchs's confession especially alarmed the public: that he had an American contact known as "Raymond." To be frank, the Soviets probably should have liquidated Gold at this point, but for some reason they hesitated. In truth, Gold nearly did the job for them. After Fuchs confessed, Gold panicked and considered killing himself with sleeping pills. His old friend Tom Black, who'd first pushed him into espionage, had to talk him out of it.

Meanwhile, the FBI began what one agent called a "raging monster of a quest" to find Raymond. The Bureau investigated 1,500 suspects, with a dozen agents working full-time on the case and sixty more working part-time. Based on Raymond's background in chemistry, they requested information on 75,000 combustible material permits issued in New York in 1945. They even sent agents to bus stations across New Mexico to ask if anyone remembered a husky fellow holding an envelope five years earlier.

The FBI finally caught a break when agents (illegally) broke into Abe Brothman's lab in New York and found several papers Gold had written on thermal diffusion. This excited them, because Fuchs had worked on *gaseous* diffusion for the Manhattan Project. In truth, the two processes have little in common, but no one at the FBI grasped that. (Sometimes it's better to be lucky than smart.) They cracked open Gold's old case file and used this and other clues to link him to Raymond.

On May 15, two agents in Philadelphia interrupted Gold at his lab around 5 p.m., and he agreed to accompany them downtown to answer some questions. They kept him there until 11 o'clock. Gold never cracked, but he returned to the lab afterward to finish his experiments—and no doubt also to calm himself.

Feeling he had no choice, Gold submitted to more hours of interrogation that weekend. When the agents asked if he'd visited New Mexico during the war, he denied ever being west of the Mississippi. When they put a picture of Fuchs before him, he nodded and said, "That's the English spy." But he claimed to recognize Fuchs only from magazines. After further pressure, he finally agreed to "settle the matter" by letting the FBI search his home—but not until Monday, when his brother and father would be absent. The agents couldn't have been happy with this; it would give Gold time to purge anything incriminating. Still, lacking a warrant, they agreed to the delay.

Incredibly, though, Gold didn't purge a thing. He headed to the lab instead. He had a few experiments running on ways to detect potassium in the body, and he couldn't bear to leave them unfinished. He then had one last dinner with his brother and father on Sunday night—"to salvage a few more precious hours" of normalcy, as he put it.

Only at 5 a.m. on Monday morning did the purge begin. Digging through his room, Gold found old train tickets, letters from Soviet agents, draft reports of interviews, and more. He scrambled to flush some items down the toilet, and buried others in the rubbish bin in the cellar.

He'd just finished up when two FBI agents knocked around 8 a.m. Wearing pajamas, Gold led them up to his room, which they began to ransack, pawing through drawers and pulling things off his shelves. There were old schoolboy notes, lab books, chemistry and physics texts. He also had volumes of poetry and several potboiler mysteries. Gold cringed when he spotted the Thomas Mann novel sitting there, the one he'd used when trying to track down a missing Fuchs. But, ever the "disciplined athlete," Gold kept cool and chatted with them as they worked.

Around 10 a.m., one agent pulled down a favorite book of Gold's—a well-thumbed copy of *Principles of Chemical Engineering*. It must have brought a smile to his face, the old standby. But of all the books he owned, this one betrayed him. As the agent opened it, a tan street map slipped out, titled "New Mexico, Land of Enchantment." Gold had grabbed it at the museum before his meeting with Fuchs.

The agent picked the map up and turned to Gold. "So, you were never west of the Mississippi . . ."

Gold all but collapsed onto a chair. He asked for a minute to think, then bummed a cigarette, which he normally hated. At this point, Gold probably still could have walked. The FBI had no hard evidence linking him to Fuchs. He even schemed up a plausible lie on the spot: a humorist he enjoyed often set stories in Santa Fe, and he could claim that he'd ordered the map as a reference. But after a decade and a half of espionage, he was simply too tired to continue—tired of lying, tired of running, tired of the burden. All he could think about was how to break the news to his brother and father.

He finally turned to the agents. "I am the man to whom Fuchs gave the information."

After his arrest, Gold vowed that he'd never rat out any of his fellow spies. *I'll accept my punishment like a man and stay quiet*, he thought. Then his brother visited him in custody. "How could you have been such a jerk?" he asked. At that moment, Gold remembered, "A good half of that mountainous mental barrier that I had erected against squealing went crashing down."

Handcuffed and accompanied by two U.S. marshals, the chemist and atomic spy Harry Gold (*center*) leaves federal court in New York on his way to jail a few months after his arrest. (Courtesy of the Library of Congress.)

Even more wrenching, Gold's father came to visit later. Samson had always been so proud of Harry—his smart son, the chemist, the one who'd gotten them through the Depression. Now he was weeping, looking frail and bewildered. "This won't affect your job at the Heart Station, will it?" he asked.

The question broke Gold's heart. "Down went another section of the mountain."

Gold soon pled guilty and spilled everything he knew. He also wrote up a 123-page document explaining his spy work in detail, then submitted to endless hours of additional questions. One agent compared interviewing Gold to "squeezing a lemon—there was always a drop or two left." Having finally unburdened himself, he felt at peace for the first time in years. His health bounced back as well and he quickly lost dozens of pounds.*

The FBI opened forty-nine separate espionage cases based on Gold's testimony. But history remembers one of them above all—the Rosenberg case. Gold couldn't recall the name of Ethel Rosenberg's brother, David Greenglass, but did remember his wife's name, Ruth, and the street their apartment was on in Albuquerque. Greenglass's furlough also coincided with the time frame Gold provided for their rendezvous. When caught, Greenglass confessed everything and claimed that he'd been pushed into espionage by Ethel and her husband Julius.

Ultimately, Greenglass was the one who doomed the Rosenbergs to the electric chair, and the press savaged him for turning against his sister. But Gold's reputation took a beating, too, from both sides of the political spectrum. Communists smeared him as a "pathological liar" and lonely "weakling" who puffed himself up by inventing fantastical tales. Meanwhile, anti-communists condemned him as a stooge who'd betrayed his country. These accusers included the prosecutors at Gold's trial, who demanded twenty-five years in jail for him, despite his extensive cooperation. The judge, who was even more anti-communist, gave him thirty. (In contrast, Klaus Fuchs—who'd actually stolen the documents—served just nine years in prison in England, after which he was deported to East Germany.†) Even Gold's fellow inmates at Lewisburg Peniten-

*While Gold felt a cloud of stress lift after his arrest, his family didn't fare so well. His father and brother got so many harassing phone calls after his arrest—most of them punctuated with anti-Jewish slurs—that they delisted their phone number. Gold had intended to combat anti-Semitism with his spying, but his exposure actually made it worse.

†As one American physicist later said, "Fuchs worked very hard for us, for this country. His trouble was that he worked very hard for the Russians, too." And it was even worse than the physicist realized. Desperate to hang onto its colonial

tiary scorned him. Thieves, rapists, hitmen—they enjoyed respect at Lewisburg. But when Gold the stool pigeon strolled over to shoot some hoops one day, every last player walked off the court. And he was lucky things didn't escalate. A few years later, a trio of prisoners at Lewisburg beat another convicted spy to death with a brick wrapped in a sock.

Once again, chemistry proved Gold's refuge. Lewisburg had an unusual prisoner health program that combined medical care for inmates with biomedical research. Nowadays, to prevent the abuses we've explored in other chapters, using inmates in medical research wouldn't fly. But ethical rules were looser back then, and Gold, a prisoner himself, jumped at the chance to return to the lab. He studied diabetes and thyroid disease at Lewisburg and volunteered to be injected with hepatitis-laced blood to help investigate a vaccine. As his crowning achievement, Gold earned a U.S. patent in 1960, from prison, for inventing a speedy blood-sugar test using a chemical called indigo disulfonate.

In his spare time, Gold did shifts in the sick ward adjacent to the lab and helped nurse inmates back to health. This went a long way toward rehabilitating him in their eyes. Indeed, Gold proved a model prisoner overall, and in April 1966, after serving 16 years, he earned parole. Once again, his case became national news. On the day of his release, his lawyer came to pick him up and was terrified to hear an uproar inside. It sounded like a riot. In reality, it was Gold's fellow inmates, cheering. After his years of selfless dedication, they were giving him a roaring sendoff.

Gold had spent the last several months of his sentence studying scientific textbooks in his cell at night, to catch up on new techniques since his arrest. Luckily for him, he found a lab director who

---

legacy as a world power, postwar Great Britain longed to be one of the first nuclear powers in the world. So Klaus Fuchs actually stole documents from Los Alamos for the British as well. Overall, then, Fuchs played a key role in producing atomic bombs for three different nations. Physicist Hans Bethe once said of Fuchs, "He's the only physicist I know who truly changed history."

believed in second chances and hired him at another hospital in Philadelphia. Gold settled into a quiet life there doing hematology and microbiology and mentoring young scientists, a kindly uncle figure. The only time the façade cracked was when someone mentioned the Rosenberg case. Once, during a news broadcast, a picture of David Greenglass flashed onto the screen. To his coworkers' shock, Gold erupted and screamed at them to turn it off.

Eventually Gold's health declined due to a weak heart. (He had a congenital defect that was probably exacerbated by the hepatitis-laced blood he'd received in prison.) In August 1972, he underwent a risky surgery to replace a valve, and died on the operating table at age sixty-one. People in his lab cried when they heard the news.

Gold had once hoped to make a name for himself as a scientist after prison: "Sometime in the future I shall be able to make far greater amends than I have done to date. And this restitution shall not consist in forming and giving evidence to the FBI . . . [but] in the field of medical research." It was another fantasy. Gold is still vastly better known for espionage than chemistry: he simply betrayed too many secrets and too many people. But unlike most communist spies, Gold had higher ideals than politics. Deep down, he was a chemist first and foremost—a man so obsessed with science that he preferred wrapping up some experiments to purging his files and saving his own neck.

Unfortunately, other scientists did let Cold War politics corrupt their integrity. Not all of them were communists like Fuchs and Lysenko, either: Fear of the Soviet Red Menace infected scientists on the free side of the Iron Curtain, too. In particular, a group of psychologists working with the CIA and U.S. military devised a system of abusive interrogation techniques that tortured dozens of innocent subjects and led to several untimely deaths. And from this milieu would emerge perhaps the most notorious terrorist in American history.

10

TORTURE: THE WHITE WHALE

Imagine it's 1960, in Cambridge, Massachusetts. Two young men are sitting in a harshly lit lab, being watched by researchers. One of the young men is smiling wickedly; the other is squirming, getting more and more agitated each minute. They're classmates at Harvard, and they're engaged in a debate about their philosophies on life. Teenagers often have strong opinions, and the agitated one—whom the researchers have dubbed "Lawful"—is especially strident.

As things get heated, their voices rise. Lawful's heart starts racing, and he squints under the hot lights. When he signed up for the study, he was told that the debates would be friendly, but his classmate keeps being abusive—mocking Lawful's points instead of critiquing them in a logical way. Today, the classmate goes further than ever. He looks Lawful up and down and sneers, *And another thing: your beard looks stupid.*

Lawful blinks, stunned. This isn't how debates work. You criticize a person's *arguments,* not the person himself. He flushes red, nearly growling, and leans forward in a surge of adrenaline. He's been coming to these "debates" for months now, and the heart-rate monitor strapped to his chest has never spiked like this.

Lawful would be even angrier if he knew the truth: that his interlocutor isn't a Harvard student at all. He's a hotshot young lawyer who's been coached to fight dirty and make ad hominem attacks. Lawful is one of twenty-two Harvard juniors suffering this abuse week after week as part of a psychological study. But none of the others show such intense reactions. Perhaps that's why the lawyer loves provoking him.

Meanwhile, the scientist studying Lawful watches the exchange through the remove of a two-way mirror. Lawful is focused on debating, but every so often he can see murky movements behind the mirror, like glimpses of life underwater. There, arms crossed, stands Henry Murray, a Harvard psychologist whose work on interrogations has piqued the interest of the CIA.

An observer once described Murray as "urbane, witty, and attentive," but also "so charming as to be suspect." He masterminded this entire setup—the noir-detective lights, the mirrors, the heart-rate monitors. In a later paper on the study, Murray will admit that the lawyer's attacks were "vehement, sweeping, and personally abusive," but in truth that's exactly what he wants. He wants to see Lawful crumble. Murray records each interaction on film as well. He does so partly to study the students for signs of frustration—their tics and frowns and furrows. From time to time, however, he also forces the students to watch the films, so they can *see* themselves frothing and sputtering on camera. It's a way to wring a little more humiliation out of each session.

And no one would be more humiliated than Lawful. The boy was undeniably brilliant (his IQ was 167), but tests revealed him as the most alienated student in the study. Indeed, Murray took special interest in the boy for that reason, and had personally bestowed the nickname *Lawful* in a mocking reference to how uptight he was. The boy's real name was Theodore Kaczynski. The world would later know him as the Unabomber.

Murray and Kaczynski came from entirely different worlds, one blue-collar, one blue-blood.

Murray was the blue blood. He grew up in an elegant brownstone in Manhattan that sat where Rockefeller Plaza does today. As an adult, he proudly listed his ancestor—the Earl of Dunmore, first governor of New York colony—on his resume.

Naturally, Murray attended Harvard and captained the crew team there. But despite becoming one of the most eminent psychologists of his generation, the field didn't exactly set him on fire at first. He majored in history at Harvard and enrolled in just one psychology course before getting bored and dropping it. (He later joked that he never again set foot in a psychology class until he had to teach one.) He then earned a medical degree from Columbia University but abandoned plans to enter surgery due to poor dexterity. (A botched surgery as a child had left one eye adrift and interfered with his hand-eye coordination.) He ended up pursuing a Ph.D. in biochemistry at Cambridge University instead, but he hardly distinguished himself there. As one critic commented, "There is nothing in his education . . . or subsequently in medical school to suggest that he was suited for anything but the life of a patrician clubman."

He finally found his calling when, in 1923, at age thirty, he came across a book by the Swiss psychoanalyst Carl Jung at a used bookstore in New York. He started reading it right there in the aisle and, electrified, skipped work the next two days to finish it. He was soon making plans to visit Jung in Switzerland to study with him.

Truth be told, Murray also sought out Jung for selfish reasons. Murray's wife Josephine was his emotional rock, a steadying presence in his life. Unfortunately she didn't excite him, especially sexually. The woman who did excite him was his mistress, the artist Christiana Morgan, despite—or perhaps because of—the fact that Morgan was flamboyant and unstable. Murray was smitten with her but couldn't bear to abandon Josephine, and he kept going round and round about what to do. So he sought Jung's opinion on the matter. Coincidentally, Jung also had a wife and a mistress, and while he listened to Murray agonize over the matter, Jung interrupted and suggested that maybe Murray didn't need to choose. He could do as Jung did, and keep both women around. After all, they were both dynamic, creative men. How could they be expected to settle on just one woman?

Naturally, this humiliated Murray's wife (she thought Jung a "dirty old man," one historian noted), but Murray brushed her concerns aside and embraced Jung's advice. Like many people drawn to psychology, the field attracted him in part because he had demons of his own (beyond his messy love life, he was a longtime amphetamine addict), and he was mightily impressed with how Jung had resolved his dilemma. Psychology seemed noble to him, a way to help people with their toughest issues, and from that moment forward Murray dedicated his life to picking apart the workings of the human mind.

Over the next few decades, Murray pioneered a new approach to psychology. Psychology at the time was split into two warring camps. One consisted of psychoanalysts like Jung, who explored the vast, murky waters of the subconscious. But many scientists criticized psychoanalysts, not unjustly, for lacking rigor. Meanwhile, the other camp reveled in rigor, almost to a fault. These psychologists would zero in on some scrap of the sensory or nervous system in an animal, present different stimuli, and measure the animal's reactions with stopwatches and electrical gizmos. It was all reflexes and rats running through mazes—anything but hard data was considered mushy.

Harvard psychologist Henry Murray, who ran an abusive psychology experiment on several students, including Theodore Kaczynski, the future Unabomber. (Courtesy of Harvard University Archives.)

Murray appreciated data-driven science—he'd come out of biochemistry, after all—but he also longed for something richer. Individual personalities fascinated him, and he wanted to study human beings the way that novelists did, in all their messy particulars. In fact, he eventually abandoned Jung as his role model and turned toward Herman Melville, whom Murray considered the true discoverer of the subconscious. (Although tellingly, when Murray read *Moby-Dick,* he identified not with Ishmael or Queequeg, but with the megalomaniacal Captain Ahab.) In the end, Murray split the difference, taking a novel, data-driven approach to studying personalities.

This middle path pleased no one, of course, but Murray's wealth shielded him from the consequences of being an iconoclast. Despite running counter to both main currents of psychology then, he used his social connections to win an appointment at Harvard. Resentful colleagues later denied him tenure, but rather than dump him, university brass spun Murray off into his own fiefdom, the Harvard Psychology Clinic, on whose door he painted a white whale. Whenever the clinic faced a budget shortfall, Murray simply pulled out his checkbook and covered the costs himself.

During World War II, Murray also used his social connections to win special assignments from the Office of Strategic Services (OSS), the precursor to the CIA. Some of this work looks awfully dodgy by today's standards. One project involved piecing together a psychological profile of Adolf Hitler, to predict his behavior during the war and suggest ways to influence him. Murray described Hitler as a blend of "artist and gangster" and "a compound, say, of Lord Byron and Al Capone." He even speculated on the erotic adventures of the Führer: "Rumor has it that Hitler's sexual life . . . demands a unique performance on the part of the women, the exact nature of which is a state secret." However titillating, this assessment has zero basis in fact.

Other work for the OSS proved more worthwhile. To help the agency manage a flood of job applicants, Murray helped devise tests to sort people into types and determine what jobs they were suited for. If you've ever taken a personality test, you're familiar with this work. Murray also helped devise a system to probe how well applicants could lie under pressure, read other people's weaknesses, and stand up to interrogation. In other words, he devised a system to find good spies.

More than anything else he did during the war, this spy work fascinated Murray—especially the dynamic between interrogators and captives, which seemed rich with drama. So, when the war ended, he decided to study interrogations systematically at Harvard. It just so happened that the OSS's successor, the CIA, shared his fascination and saw this work as a way to gain an edge during the Cold War.

In the worldwide battle between communism and democracy, communism seemed poised to crush its rival in the late 1940s. Joseph Stalin and the Soviet Union had conquered all of Eastern Europe, and Mao Zedong had seized control of the world's most populous nation, China. War nearly broke out in Europe over Berlin, and did break out in Asia over Korea. The threat of atomic weapons then ratcheted the tension higher still.

However scary, though, atomic bombs were merely external threats. Even more frightening to many Americans was the prospect of their minds being invaded—of communists conquering their souls from within. Thousands of POWs in Korea signed declarations in which they "confessed" to crimes they never committed, such as using germ warfare against enemy troops. Similarly, during the Soviet Union's notorious show trials in Eastern Europe, defendants confessed to absurd deeds—and did so with slurred words and glazed, zombie-like looks. CIA analysts reviewed this evidence and took a big but understandable leap. They concluded that communist interrogators had discovered a surefire method of brainwashing people: a psychological superweapon that allowed them to crack open people's minds and turn them into "Manchurian candidates" that would do their evil bidding.

In truth, the communists had no such ability. What they did have was torture.

Sometimes communist interrogators employed "mild" torture techniques like isolation and sleep deprivation. When that didn't work, they either beat their captives silly or got creative and did things like inject chemicals to induce seizures, or wrap people in wet canvas that shrank and crushed them as it dried. To be sure, these methods were probably useless for gathering intelligence, based on what we know about the science and psychology of torture. There's of course never been a genuinely scientific study on whether torture works. Such a study would require splitting a pool of volunteers into two groups, one of which gets tortured to reveal their innermost secrets and one of which gets interrogated through humane means, and then comparing the results. Even compared to other cases in this book, a study like that would be crazily unethical, and in the absence of a rigorous study, we can't draw any firm conclusions about the efficacy of torture. Still, the best science out there does cast doubt on its reliability. Even mild stress compromises our ability to recall information, and there are fewer things more stressful than torture. Moreover, studies have shown again and again that people under duress will confess to all sorts of ludicrous things simply to find relief from torment. However crazy it seems from the outside, false confessions happen all the time.

If you want reliable intelligence, then, there are better ways to acquire it than torture.* But of course Soviet and Chinese communists

---

*In case you're wondering, there are methods out there for getting useful, reliable information during interrogations.

In the rock-'em-sock-'em 1930s, cops would often coerce confessions by dunking suspects in water or dangling them out of windows. Eventually, these techniques were condemned as barbaric, and replaced by supposedly advanced psychological methods involving harsh lights, isolation, and the good cop/bad cop dynamic. Unfortunately, despite their scientific veneer, these new methods didn't work very well and led to thousands of false confessions over the years. Equally bad, even when police officers collared the right suspect, their aggressive questioning often caused him to clam up and stop talking, which stymied investigations.

weren't necessarily seeking reliable information; they often *wanted* false confessions, which they then could use as propaganda. In that narrow, cynical sense, torture does "work," works brilliantly.

Unfortunately, such subtleties were lost on CIA analysts. They were terrified that the communists had discovered the psychological equivalent of the atomic bomb, which would give them a decisive edge in the Cold War. As a result, the CIA instituted a crash program called MK-ULTRA in 1953 to narrow the "mind-control gap." "MK" indicated that the project fell under the scope of the CIA's Technical Services Staff, while "ULTRA" was reportedly a nod to the ULTRA program during World War II, which involved cracking German military codes. The implication was that defeating the Soviet Union in this arena was every bit as crucial for the survival of democracy as defeating Nazi tanks and submarines had been. We had to develop new methods for hacking the human brain as quickly as possible, and do to the commies what the commies had been doing to us.

Under MK-ULTRA, any method for probing the mind was considered fair game. Grim, humorless CIA analysts teamed up with fortune-tellers and occultists to study hypnosis, telepathy, clairvoyance, and other voodoo, all on the off-off-chance that some of it wasn't baloney. MK-ULTRA is best known today for its reckless deployment of LSD, which agents hoped would prove useful as

---

Nowadays, the best interrogation methods focus less on seeking confessions and more on getting bad guys to incriminate themselves by talking too much. The thinking is that the more people blab, the more likely they are to contradict themselves or else spill details about their movements and alibis that detectives can check into and trip them up with. Cops can also use tricks like asking people to tell a story backward, or having them sketch an unrelated picture while going over their story, since doing so increases the "cognitive load" on them and makes it harder to keep their lies straight.

Admittedly, these "soft" methods don't satisfy our need for revenge—the sense that bad guys deserve rough treatment. But innocent people get arrested on false charges all the time. And if your goal isn't to enact revenge as much to see bad guys actually convicted and locked away, then helping them relax and letting them blab until they screw up is far more effective.

a truth serum. (The army did similar work with PCP and mescaline.) To be fair, the agents did at least dope themselves first. They'd spike their colleagues' wine or cigarettes at parties, then follow them around town as they went screaming mad, convinced that "monsters" (really passing automobiles) were trying to devour them. But it wasn't long before agents started doping strangers. Sometimes they used drug-coated swizzle sticks; other times they hired magicians to teach them sleight-of-hand, so they could slip mickeys into drinks in bars and brothels. (One particularly creepy agent ran a bordello in San Francisco where he'd watch LSD-fueled sex from behind a two-way mirror—and do so while sitting on a toilet with a pitcher of martinis in hand.) In fact, LSD might be little more than a laboratory curiosity today if the CIA hadn't tested it so widely. It's one of the twentieth century's great ironies that this straitlaced, ultraconservative agency inadvertently spawned the counterculture drug movement of the 1960s. The Grateful Dead were as much a child of the Cold War as fallout shelters were.

MK-ULTRA was eventually shuttered for lack of results, and a later director of the CIA ordered all files related to it destroyed. As a result, the full extent of the program remains murky. But at least 185 scientists at 86 institutions took part, and the CIA actively recruited psychologists with low moral standards. One's personnel file read, "His ethics are such that he would be completely cooperative."

Much of the research that the CIA sponsored focused on stress—both the causes of stress and the strategies people used to cope with it. In isolation, this work was worthwhile: it helped people deal with tension and anxiety in their lives. But the CIA took these results and twisted them. Once its analysts understood what caused stress, they suddenly had a blueprint for inducing it in POWs and spies, to wring supposed secrets out of them. Similarly, if analysts knew how people coped with stress, they could undermine those coping mechanisms and ratchet up the pressure even more. In its way, the strategy was fiendishly clever: academic psychologists did the work, while the CIA reaped the benefits.

It's at this point that the CIA's interests and Henry Murray's interests converged. To be clear, despite speculation among historians and conspiracy theorists, there's no hard evidence that Murray ever participated in MK-ULTRA or any other CIA program. That said, so many files were destroyed, and the work was so secretive in the first place, that the lack of paper records proves nothing. Murray certainly shared the CIA's interest in interrogations. He'd worked for the agency's predecessor on that topic, and the counterculture psychologist Timothy Leary—a colleague of Murray's at Harvard—explicitly said that Murray directed military experiments on brainwashing for OSS. So even if Murray never took a dime from the CIA, he was part of its milieu and shared its mentality.

In fact, Murray's interest in interrogation was arguably even darker and more cynical than the CIA's. However misguided, CIA analysts sincerely believed they could elicit good intelligence through torture and use that information to save the world. Murray was no doubt happy to help with that, but he mostly just wanted to brutalize people and see what the hell happened. Specifically, he theorized that by attacking people's core values and exposing those values as meaningless, he could disorient and break them—at which point, they'd be susceptible to psychological manipulation. To that end, Murray inaugurated a study in the fall of 1959 on psychological abuse in what he called "gifted college men," men like Ted Kaczynski.

If the U.S. government abused psychological science during the Cold War, it's only fair to note that the Soviet Union did as well, albeit in different ways. Rather than pursue brainwashing (which they knew was bunk), the Soviets used psychology to discredit political activists and lock them away without trials.

The system worked like this. Whenever the KGB arrested a dissident—someone who just wouldn't shut up about civil rights, or religious freedom, or abuse of power—agents would haul him or her into a mental institution. Several such institutions had KGB members on their boards, and staff psychiatrists would dutifully condemn the dissidents as insane and lock them up. The most common diagnosis was "sluggish schizophrenia," a mythical, slow-acting form of schizophrenia whose symptoms included "reform delusions," "a struggle for the truth," "perseverance," and, oddly, an affinity for abstract or surreal art. From the KGB's point of view, declaring dissidents insane had several advantages. Authorities could avoid a trial, where messy secrets might emerge. The label also smeared dissidents as mentally ill, and stigmatized their followers as equally crazy. Between the 1950s and 1980s, thousands of people disappeared into the Soviet psychiatric Gulag, where they were often drugged to keep them docile.

To be sure, some Russian psychiatrists condemned people to asylums out of mercy: their patients likely would have been executed if they hadn't been declared crazy. But most psychiatrists enthusiastically supported the KGB. Rather than the Hippocratic oath, doctors in the Soviet Union swore a special pledge to serve the Communist Party first, and they took this oath seriously. As one of them put it, doctors there "knew when to put down the stethoscope and take up the pistol." They'd also been indoctrinated to view the Soviet Union as a workers' paradise, the greatest state in world history. Defiance toward it was therefore an ipso facto sign of derangement—who else but a crazy person would oppose heaven? Some Soviet psychiatrists even mourned the rise of Mikhail Gorbachev, who seemed obviously schizophrenic to them. After all, didn't Gorbachev agitate for political reform? Didn't he natter endlessly about human rights? Only after the Soviet Union collapsed did they realize just how skewed their perspective had been.

Still, the Soviet Union was hardly alone in abusing psychology for political ends. At different points in the twentieth century, Romania, Cuba, South Africa, and Holland were all accused of similar misdeeds. The biggest culprit today is China, which has been cracking down on Falun Gong since 1999 in part by declaring it an "evil cult–induced mental disorder." One woman's case report condemned her for "flagrantly telling everyone how much she was benefiting from her practice of Falun Gong." Similarly, a male adherent was locked up for "giving people valuable presents for no reason." The nerve!

Sadly, suppressing dissent this way can be effective. Dissidents who are jailed in traditional prisons have a way of becoming martyrs. Dissidents jailed in psychiatric hospitals don't: people hesitate to voice support for someone who's deemed crazy, lest friends and loved ones start to look askance at them, too. However abusive, there is a certain logic to this system.

In contrast, the American government's abuse of psychology was anything but logical. In particular, experiments with LSD and other "truth serums" were poorly thought out and lacked any semblance of consistency. Doses ranged from minuscule hits to more than ten times a typical street dose, and there was so much secrecy surrounding the compounds that even the doctors involved often had no idea what they were injecting people with. (As one said, "We didn't know whether it was dog piss or what.") And the human cost was real. One drugged U.S. marshal robbed a bar with a gun while high. Other experiments killed people or drove them to suicide.

The Harvard research into brutal interrogation methods betrayed a similar indifference to people's welfare. To be sure, there's no evidence that Henry Murray actively set out to harm Ted Kaczynski or the other gifted young men. At the same time, Murray didn't seem to care much if he did.

As a boy growing up in Chicago, Ted Kaczynski was a strange mix of overly sensitive and overly rational. One summer, Kaczynski's father trapped a baby rabbit in their backyard using a wooden cage. The rabbit was unharmed, but as Kaczynski's brother and other boys gathered around to gawk, it naturally began to tremble. This was completely intolerable to Ted. As soon as he arrived on the scene, he began shrieking at them to let the bunny go—*let the bunny go!* He wouldn't stop screaming until his father did.

At the same time, Kaczynski could be logical to the point of cruelty. One day his brother David was flipping through some baseball cards on the grass with a friend when the friend asked who David's favorite player was. David was too embarrassed to admit that he knew nothing about baseball, so he peeked down at the cards and blurted out the first name he saw. To his relief, the friend also liked the player. All in all, a harmless childhood fib. But when David returned home and announced to Ted who his new favorite player was, his older brother started grilling him. When had he started liking that player? Why did he like him so much? A deflated David couldn't say. "I should have known," he later sighed, "that Teddy would request a reason. All opinions . . . should be based on reasoning." Another time, David said to Ted, "Aren't we lucky we have the best parents in the world?" Kaczynski shot back, "You can't prove that."

Still, David more or less worshipped Ted, who was equally brilliant in both math and music. As David once said, "I couldn't tell if my brother was going to be the next Einstein or the next Bach."

Ted's parents were less optimistic about the boy's future; there was just something off about him. As a child, Kaczynski refused to hug people back—he always squirmed whenever people wrapped their arms around him. He also struggled to make friends. Even when his mother bribed other children with lemonade and cookies to come over and play with him, Kaczynski showed little interest. Things got even worse when he took an IQ test in fifth grade. He scored 167, and his principal recommended skipping him ahead a year in school. His parents didn't realize that uprooting an already small and awkward boy and planting him among older children would further isolate him, and they did as the principal said.

Now, plenty of loners grow up to be healthy, functioning adults. And Kaczynski did make a few friends in school and did take part in social activities like band (he played trombone). He wasn't a complete loner, then. But his parents, Turk and Wanda, fretted constantly over his lack of popularity and even bullied him about it, calling him "sick" and "immature" and "emotionally disturbed" for not dating or joining boy scouts. The words scarred Ted. Turk and Wanda then compounded their earlier error by having him skip his junior year of high school as well. Because he had a May birthday, he was now two-plus years younger than most of his classmates.*

At age fifteen, Kaczynski won admission to Harvard. It should have been a proud moment, but his band teacher begged Turk not to let the boy enroll there. However smart, the teacher argued, Ted simply wasn't ready emotionally for such a pressure cooker of a school. Equally bad, Kaczynski would never fit in there socially. Turk made sausages for a living at his uncle's shop near the Chicago stockyards, a livelihood that your typical Harvard lad—the scion of a banker or senator—would scorn. The teacher argued that nearby Oberlin College, which also had a strong music program, seemed a better fit. Turk wouldn't hear of it. Despite his menial job, he was a proud man who read voraciously, and he wanted the best for his son. Harvard it was.

---

*During high school Kaczynski actually helped make a bomb once, but at the time it didn't register as a big deal. The whole thing wasn't even his idea. He happened to have a classmate who was fascinated with bombs. Being a good chemistry student, Ted knew how to produce an explosion by mixing ammonia and iodine together; if you then touch the mixture—even with a feather—voilà, it's a crude bomb. When the classmate heard this, he begged Ted to teach him how.

Now, Ted probably shouldn't have relayed this information. But he did so in a sad attempt to impress the other boy, a popular wrestler, and make a friend. It didn't work. The bomb did, unfortunately, and when the boy deployed it during chemistry class one day, it blew out two windows and one girl temporarily lost her hearing. Thankfully, everyone else in the vicinity escaped unharmed. The school's principal saw the incident for what it was—a knucklehead stunt—and suspended Kaczynski for a day, then forgot about the whole matter. Only decades later, in retrospect, did it seem ominous.

To be fair, in sending Kaczynski to Harvard, his parents had more in mind than just their own pride. They'd all but given up on Ted fitting in at his high school. As psychologists have documented, people with genius IQs often have trouble relating to peers, since their brains simply work differently. (Kaczynski refused to play with the boys in his neighborhood in part because he found them moronic.) But at Harvard, his parents reasoned, Ted would finally meet people on his own level. He would make friends at last and settle into a normal life.

If only. For Kaczynski's freshman year, a well-meaning dean placed him in a small dorm full of similar students: high achievers from atypical Harvard families. In theory, this common background would help them make friends with each other. In practice, the building contained mostly other misfits who also had trouble socializing. It turned into what one historian called a "ghetto for grinds." Again, Kaczynski did make a few friends at Harvard; he wasn't completely alone. Harvard, though, was no less forgiving of poverty in the 1950s than it had been in John White Webster's day

Theodore Kaczynski, the future Unabomber, as a young man (*left*). Family photo of, respectively, Kaczynski, his father, Turk, and his brother, David (*right*). (Courtesy of the U.S. Marshals.)

in the mid-1800s. Appearances mattered, and Kaczynski's awkward manner and threadbare clothing marked him as a loser.

Kaczynski first made the acquaintance of Henry Murray during his sophomore year. At the time Murray was using a psychology course on campus to recruit students for his study on abusive interrogation. Not that he described it that way. On the recruitment fliers he hung, he styled the research, blandly, as a chance "to contribute to the solution of certain psychological problems." Kaczynski didn't take the course, so it's not clear how he and Murray met. Perhaps Kaczynski saw a flier in passing and volunteered, or perhaps Murray became aware of the young grind and recruited him. Regardless, a preliminary screening identified "Lawful" as the most alienated youth in the cohort, which apparently gave Murray no pause.

Because Kaczynski was only seventeen then, a minor, Murray had to write to his parents in Chicago and get their permission to enroll him in the study. Poignantly, Wanda Kaczynski didn't quite grasp what she was getting her son into. All she understood was that Teddy still wasn't making friends, even at Harvard, and that maybe the nice psychologist in the letter could help her boy with his troubles. She signed him right up.

The other students signed their own consent forms, which were every bit as deceitful as the fliers. Murray first asked them to write up their "personal philosophy of life . . . the major guiding principles in accord with which you live or hope to live." He then claimed that they'd be engaging in a friendly debate about these philosophies with another student—failing to mention that the "student" was really a lawyer he'd coached to be aggressive and cruel. As one of the interrogees said, "I remember being shocked by the severity of the attack."

Lying to his subjects was already a violation of the Nuremberg Code for ethical research, but Murray wasn't done there. He also concealed the true purpose of the experiments, and no wonder: The goal was essentially to shatter the young men. In fact, while the Code aims to minimize suffering during research, the whole point of Murray's study was to *produce* anguish. Finally, while Murray did allow the students to drop out of the study if they wanted, he employed every bit of his considerable charm and authority to keep them enrolled—effectively coercing them by claiming that their leaving the study would destroy it.

True to his nickname, Lawful proved one of the more dutiful subjects, enduring more than 200 hours of abuse over three years—including one session, per the imagined scene above, where the lawyer made fun of his fledgling beard. Kaczynski later called the study "the worst experience of my life," but he persisted in coming back week after week for several reasons. One was cussedness. "I wanted to prove I could take it," he once said, "that I couldn't be broken." He also did so, one historian speculated, because Murray paid the students and the blue-collar Kaczynski needed the cash.

Depending on your point of view, what Murray did to Kaczynski might not rise to the level of torture. After all, he never laid a finger on the boy, nor threatened him or his loved ones. But there's no question Kaczynski suffered at Murray's hands, suffered terribly—both his later recollections and his reactions at the time (e.g., his racing heart) testify to that. What's more, if we can extrapolate from the experience of torture victims to Kaczynski's experience, the type of suffering he endured must have been particularly painful. Those who suffer for a cause they believe in tend to bounce back more readily after their plight. This is referred to as the Joan of Arc phenomenon: their suffering is noble, and their torture anneals them. In contrast, those picked out at random for abuse, or those caught up in a crusade they didn't choose, tend to suffer more deeply and often struggle to rebound. There's no story they can tell themselves, no noble cause that acts as a balm. Torture degrades them.

Now, it's too reductive to say that the Murray experiment "made" Kaczynski into the Unabomber. After all, twenty-one other students endured the same abuse, and none of them went on to build a shack in Montana and mail bombs to people. As Murray would have been the first to point out, individuals are complicated and idiosyncratic, and causes and effects are rarely straightforward. Along those same lines, it's also too simplistic to ascribe Kaczynski's violence to "bad genes" or growing up in a "bad home."

It is, however, plausible to talk about a *combination* of bad genes and bad experiences as a trigger for violence. For example, consider the *MAOA* gene on the X chromosome. This gene produces a protein that helps break down neurotransmitters in the brain. Different versions of the gene break down neurotransmitters at different rates, and the presence or absence of different neurotransmitters can affect our thoughts, emotions, and behaviors. This is relevant because people born with certain versions of *MAOA* are far more likely to be violent and show antisocial behavior—but *only* if they suffer abuse or neglect as children. If they don't suffer abuse or neglect, they're normal. You need both the bad gene and the bad experience to see the effect.

Now, in the absence of genetic testing, we can only speculate about Kaczynski's DNA. But he was clearly high-strung as a child, even hysterical (recall the rabbit). On top of that, he grew up in a dysfunctional home where his parents called him "sick" for not making friends and where he faced impossible-to-reconcile pressures to both excel academically and yet be perfectly normal socially. Frankly, his genius probably didn't help him, either. Many gifted people are psychologically fragile. They're like orchids, which can blossom if cultivated with care, but often wilt in adverse circumstances. If Kaczynski was indeed vulnerable to mental illness, his home and school life probably compounded his problems.

Then he met Murray. Again, you might or might not buy the comparison to torture, but the most effective way to break detainees under interrogation is clear: you isolate them, you stress them, and you keep it up over time. The sixteen-year-old Kaczynski was already isolated when he arrived at Harvard, and things only got worse in the ghetto for grinds. Murray's study then heaped unnecessary stress onto him, stress that the stubborn Kaczynski endured for three years. In fact, prolonged stress can cause permanent changes to the brain, atrophying some parts and putting others—like the circuits that process anger and fear—on a hair trigger. This doesn't mean that we can pin the blame on Murray for Kaczynski's murders, any more than we can pin the blame on the kids who wouldn't play games with Teddy growing up. But adolescence is a formative time, and Murray's unethical, years-long experiment could well have pushed an already fragile young man over the edge.

With all this in mind, it's instructive to compare Kaczynski to his younger brother. If anything, David was even more alienated than Ted as a young man. They both attended Ivy League schools (David went to Columbia) and then retreated to the wilderness to spurn society. In fact, David lived far more primitively than Ted ever did in his Montana cabin. At least Ted had walls and a proper roof. David simply dug a grave-shaped hole in the Texas desert and pulled a tin sheet over the top at night. Biblical hermits lived more luxuriously. Yet, despite feeling equally alienated from society, and despite growing up in the same home environment and sharing many of the same genes, David's orchid survived—perhaps because he never endured something like the Murray experiment. After eight years, David finally quit the Texas desert and married his high-school crush. Even more telling, while Ted scorned traditional morality as a bourgeois tool of mind control, David had a strong enough moral compass to turn his brother in* after reading the so-called Unabomber manifesto in a newspaper years later.

---

*To be accurate, it wasn't David as much as his wife Linda who cracked the case. After reading the manifesto, Linda asked David if it were possible that Ted (whom she'd never met) might be the Unabomber, given their mutual contempt for industrial society. David rejected the suggestion at first, but eventually admitted the idea might hold water, based on several clues he couldn't ignore.

For instance, David noticed that some of the bombs had been detonated shortly after the Kaczynski family had sent Ted money. Ted also knew carpentry (several bombs had wooden components), and he'd lived in several cities where the bombs killed people. Finally, David recognized certain phrases in the manifesto that his brother used in letters (e.g., "cool-headed logicians"), as well as his brother's idiosyncratic spelling (e.g., "analyse," "wilfully"). Along those same lines, one of the

To be sure, Kaczynski himself downplays the link between the Murray study and his crimes, a connection he sees as simplistic and sensationalistic. (He blames his parents for his troubles, which seems equally reductive.) But Kaczynski's own words betray just how formative the experiment was. His now-famous manifesto focuses primarily on how modern technology degrades and debases the human spirit; its first line reads, "The Industrial Revolution and its consequences have been a disaster for the human race." But he also takes repeated swipes at psychology and psychologists, mentioning them dozens of times. During his trial he told his lawyer, "I am bitterly opposed to the development of a science of the human mind." He then wrote the judge, "I do not believe that science has any business probing the workings of the human mind." Even back at Harvard, Kaczynski was already having trouble sleeping, and for years afterward he was wracked with nightmares about psychologists "trying to convince me that I was 'sick' or . . . trying to control my mind through psychological techniques."

Coming from anyone else, this would sound like tinfoil-hat paranoia. *The CIA is invading my brain, man.* But a psychologist with ties to the intelligence world really did experiment on Kaczynski, and really did try to break his mind.

The Murray-Unabomber case is unusual in the annals of unethical science in that the victim ended up committing sins far more grievous than the perpetrator.

FBI agents later noticed that both Kaczynski's letters and the manifesto contained the phrase "can't eat your cake and have it too," rather than the more common "can't have your cake and eat it too." If you think about it, the second one is illogical, since you can indeed have your cake, wait a while, and then eat it. What's logical is the first one—you cannot eat your cake and *then* have it. Ted had of course thought this through and insisted, to his doom, on using the correct version.

There are few studies out there on geniuses who commit crimes, but one strong predictor of trouble is a mismatch in the brain between IQ and EF, the so-called executive function. Executive function is located primarily in the frontal lobes (the lobes that were removed in the chimpanzees Becky and Lucy). It helps us manage our impulses, make decisions, and impose self-control, among other things. As one psychologist put it, "IQ operates like the raw horsepower of a car engine, while EF . . . operates like the transmission, directing the power" toward useful ends. But if someone's IQ greatly outstrips their EF, then you essentially have a drag-racer without a steering wheel: the car can easily careen out of control and send the owner flying off the path of acceptable behavior. What's more, while geniuses typically commit crimes for the same basic reasons as the rest of us schlubs (greed, jealousy, etc.), their crimes are often more complex and require more sophisticated planning. Kaczynski spent years designing and testing his bombs, keeping meticulous encrypted notes* on each "experiment." He also took elaborate steps to conceal his tracks, like soaking every component in saltwater and soybean oil to remove fingerprints.

Finally, criminal geniuses seem especially prone to nihilism—a rejection of conventional morality in the belief that life is meaningless. Kaczynski certainly subscribed to this idea, as his journal made clear. (For example, "Morality is simply one of the psychological

*Kaczynski's encryption system was genius. It started with a list of numeric substitutions: 4 = THE, 18 = BUT, 1 = the present tense forms of TO BE, 2 = the past tense forms of TO BE, and so on. This list included individual letters, too, with 39 = A, 40 = B, et cetera. But he also introduced some twists. Both 62 and 63 equaled S, for instance, and 45, 46, and 47 all equaled E, to throw off any attempts to use letter-frequency counts in deciphering the text. He even used different letters for a voiced and unvoiced "TH," and lumped all forms of ME, MY, and MINE together under one number. Even more devilishly, he intentionally misspelled words, included strings of nonsense sometimes, and swapped in German and Spanish words (he spoke both languages) whenever it suited him. All these ploys—and the addition of other encryption techniques—would have made the codes nearly impossible to decipher without supercomputers and dedicated effort.

tools by which society controls people's behavior.") Perhaps he would have come to this conclusion on his own, no matter what happened to him during his life; plenty of people do. At the same time, it might not be a coincidence that the hotshot lawyer in Murray's study had been specifically coached to attack the students' values and expose them as meaningless. If you want to make a nihilist, that's a pretty good start.

Like many lawbreakers, genius and otherwise, Kaczynski began with petty crimes. After moving to his cabin in Montana, and finding the area far less isolated than he hoped, he began destroying nearby logging equipment. He also chopped his way into a neighbor's luxury cabin and trashed the place, then smashed the man's motorcycle and snowmobiles for good measure. Kaczynski soon graduated to more serious offenses like shooting rifles at passing helicopters and stringing up wires to snag snowmobiles. Eventually he started building bombs and either mailing them to strangers or leaving them in well-trafficked public places. He didn't quite pick his targets at random, but he didn't put much thought into them, either. Nor did he delude himself that a few explosions would take down the whole "system" of "psychological controls" that he despised. He was simply angry and lashing out, hoping to kill.

Sadly, he succeeded. In all, Kaczynski set off sixteen increasingly sophisticated bombs between 1978 and 1995, maiming several people and killing three. The FBI dubbed him the Unabomber because he targeted universities and airlines, and the manhunt for Kaczynski was the longest and most expensive case in FBI history, outstripping even the raging monster of a quest to find Harry Gold.

Kaczynski took advantage of his notoriety to promote his theories about the corrosive effects of industrialization on the human spirit. He even promised to stop killing people if the *New York Times, Washington Post,* or *Scientific American* published his manifesto. (Oddly, *Penthouse* volunteered to publish it as well, but Kaczynski reserved the right to kill one last person if it appeared only there, given its less salubrious reputation.) The *Post* did publish the manifesto, but like a good nihilist, Kaczynski had no intention of keeping his promise. Just before his arrest, he was rasping bars of aluminum into powder for yet another bomb.

Despite his unkempt hair and clothes, Kaczynski's cabin was actually pretty tidy when the FBI raided it. Agents had been staking out the place for weeks, and on the morning of April 3, 1996, several of them began crawling through a dry creek bed toward it; others hid out in the surrounding woods. A local forest ranger whom Kaczynski knew then approached the cabin with two undercover agents, and pretended to be having an argument about the exact boundary of Kaczynski's property. "Hey, Ted," the ranger finally hollered. "Can you come out here and show us where it is?"

Inside the Unabomber's cabin in Montana. Contrary to media reports, it was fairly neat and tidy. (Courtesy of the FBI.)

Kaczynski popped his head out. He was habitually wary, but said, "Sure. Just let me go back in and grab my jacket."

As soon as he turned his back, one of the undercover agents sprang, pouncing on Kaczynski and wrenching his wrists into handcuffs. The other agents then burst out of the woods, swarming the cabin and keeping their eyes peeled for booby traps and other dangers.

To their amazement, the agents found that Kaczynski owned suits and ties and had dozens of classic works of literature on his shelves (Shakespeare, Twain, Orwell, Dostoyevsky, etc.). Poignantly, he also had a copy of Henry Murray's paper on the abusive interrogation of gifted young men. It ran just nine pages, and was the only thing Murray ever bothered publishing on the experiment. The study contained no profound conclusions, and no sense of regret or apology—Murray mostly talked about the subjects' heart rates. In the end, the patrician Murray blithely forgot about the young men and the abuse he'd heaped on them. Kaczynski never could.

After his capture, Kaczynski pleaded guilty to thirteen federal counts. In its illustrious history, Harvard has had just two alumni executed for capital crimes. The first was George Burroughs, for witchcraft, in 1692. The second was John White Webster from chapter four, for killing George Parkman in 1849. By pleading guilty and accepting a plea bargain, Theodore John Kaczynski narrowly avoided becoming the third.*

In trying to explain Kaczynski, the media mostly focused on his life after Harvard. He ended up earning a Ph.D. and becoming a professor of mathematics at the University of California at Berkeley. This was late 1960s, a radical decade, and "Bezerkeley" was the most radical college in the country. Calls to violence and revolt were

---

*In 2012, Kaczynski sent a cheeky update to the Harvard alumni magazine for the 50th anniversary of his graduation. Incredibly, in a monumental editorial lapse, the magazine published it. He listed his occupation as "prisoner," his address as a supermax prison in Colorado, and his "awards" as eight life sentences from the U.S. district court in California.

everywhere, and to armchair psychologists in the media the connection seemed obvious: the tumultuous atmosphere at Berkeley must have corrupted the young genius, and driven him to evil.

By his own account, however, Kaczynski was already broken by the time he arrived in California. He in fact took an academic job mostly to save up money, so he could buy land somewhere and start enacting the fantasies of murder and revenge that had crystalized back East. That he'd ended up in Berkeley was nothing but a coincidence, and he barely noticed the mayhem around him. His real troubles traced back to his twisted childhood, and the hours of abuse he endured in a building at Harvard with a white whale emblazoned on the door. Not to mention the Cold War paranoia that convinced the CIA and its allied scientists that the suffering was worth the human cost.

Still, as we've seen, science has enemies on both sides of the political aisle. In fact, while the Unabomber case revealed the danger of conservative bias run amok, our next case involved a small boy whose happiness, and ultimately life, were sacrificed to the far-left dogma of yet another rogue psychologist.

11

MALPRACTICE: SEX, POWER, AND MONEY

For no reason at all, the nurse reached into the bassinet and, rather than Brian, plucked out his twin brother, Bruce. The eight-month-old Reimer twins from Winnipeg both had phimosis, a disorder that prevented their foreskins from retracting and interfered with peeing. In that surgery-first era, the mid-1960s, their doctor recommended circumcisions to clear the condition up. The Reimer parents obliged, and the morning after they dropped the twins off at the hospital, the nurse plopped Bruce down on the operating table.

The pediatrician who normally performed circumcisions was off that day, so the duty fell to a general practitioner. He fit a bell-shaped metal instrument inside Bruce's foreskin, to stretch it out, and used a metal clamp to hold the skin in place. Rather than a knife, he then reached for an electrocauterizing needle. It was a whiz of a device, the latest thing: it sent electric pulses through a needle tip to both cut and seal flesh simultaneously, minimizing scars and bleeding. Unfortunately, this second-string doctor was apparently ignorant about the danger of mixing electricity with metal.

When he first touched the needle to Bruce's foreskin, nothing happened, so he turned the current up. Nothing happened this time, either, so he cranked it up more. This time something happened. The electricity burned through the diaphanous foreskin and flooded the metal bell beneath. From there, the current enveloped the entire penis in a sheath of heat. The anesthesiologist on hand remembered hearing a sound "like steak being seared." The room smelled like frying meat, too, and a plume of smoke leapt up from between Bruce's legs. The GP yanked the needle back, but it was already too late. By the time the emergency urologist arrived, Bruce's penis looked blanched and bloodless, like an overdone bit of pork. It felt strangely spongy, too.

Bruce's parents, Ron and Janet Reimer, soon got a call at home. The hospital wouldn't tell them what the problem was, only that they needed to hurry. A freak April blizzard had hit Winnipeg, and it took them agonizingly long to navigate the streets. But there was nothing they could do anyway. By the next day, Janet remembered, Bruce's entire penis had "blackened, and it was sort of like a little string." It dried out and crumbled off in pieces over the next few days.

The hospital never circumcised Brian, whose phimosis cleared up on its own. This was small comfort to Ron, twenty, and Janet, just nineteen, who suddenly had a mutilated baby and no idea what the hell to do.

John Money once called the penis "the mark of man's vile sexuality," and added that "the world might really be a better place for women if not only farm animals but human males were gelded at birth." If those comments startled you, well, mission accomplished. Whether it was zealous loyalty or sputtering hatred, no one was ever neutral about John Money. He always got a reaction.

Money grew up in the 1920s in a strict Christian community in New Zealand. His father beat him for minor transgressions, and his mother suffered even worse abuse. She and her sisters grew to hate the toxic men in their lives, and Money said they filled him with their biases.

At age twenty-five he fled New Zealand for graduate school in psychology at Harvard, where his colleagues included Henry Murray. He wrote his Ph.D. thesis on the psychological health of hermaphrodites (now called intersex people). Contrary to expectations—even medical textbooks tossed around words like "freak" and "misfit" and "it"—Money found that most hermaphrodites were perfectly normal and had no more psychological hang-ups than the general public. This willingness to normalize hermaphrodites (worldwide, they're as common as people with red hair) made him a hero in the intersex community.

Money soon won an appointment at the Johns Hopkins University hospital in Baltimore, where he made his most enduring contribution to sexual psychology. In human beings, several different factors contribute to sexual identity: hormones, anatomy, sexual orientation, cultural expectations, and so on. But beyond all that, Money recognized an additional factor—whether individuals *felt* masculine or feminine inside.

While masculinity and femininity usually aligned with genitals and hormones, they didn't always. You can have male genitalia but feel feminine, or vice versa, among other permutations. Money wanted a term to describe this feeling, so he looked to linguistics. Native English speakers often struggle with the fact that, in other languages, bridges are suddenly "masculine" or tables "feminine," a feature called the gender of a word. Money borrowed that term and applied it to people. In Money's schema, "sex" covered chromosomes and anatomy—physical things—while "gender" covered behaviors and feelings. In short, sex was biology, gender psychology.

"Gender" soon entered the general lexicon, and Money reveled in his fame as the coiner of the term. He then parleyed that fame into greater notoriety by taking provocative stands on social issues. Some of these stands seem quaint nowadays. People gasped when he advocated for nudism and open marriages and defended S&M. Other stands still seem reckless. During public lectures, he showed graphic slides of bestiality and feces-eating, fetishes he supported as perfectly wholesome. He also endorsed pedophilia in some cases, and got exasperated when people treated incest as a black-and-white issue. In fact, he said, stepfathers preying on stepdaughters was usually a good thing, since the mother was "glad to have [the husband] off her back."

It's hard to know whether Money believed this crap. He loved getting a rise out of people, and once described his method of theorizing about sexuality as "playing a game of science fiction." But once he carved out a position, he defended it to the death. Those who challenged him weren't just wrong or misguided, either. They were hateful, small-minded bigots hopelessly mired in the past.*

Beyond a brief marriage, Money had virtually zero personal life. People at Hopkins loathed him as a cheap S.O.B. with a volcanic temper. He made his students peel the stamps off envelopes so he could reuse them, and he raided the buffet at the hospital at night and packed the leftovers into plastic bags. Colleagues who dared point out his mistakes quickly learned to never do so again, lest they face his wrath a second time.

---

*John Money wasn't exactly known for his scrupulous scholarship, either. One example involved a tribe in Australia called the Yolngu, whom he visited in 1969. Despite spending just two weeks among them, he emerged with several sweeping pronouncements about their sexual lives. Most notably, he claimed that these adorable little primitives just loved getting nekkid and getting it on; as a result, he said, Yolngu adults had zero sexual hang-ups or neuroses, including a complete lack of pedophilia and homosexuality—both of which, he maintained, were entirely the product of Western sexual repression. Even apart from the implication that homosexuality is a neurosis, this is complete bunk. Anthropologists who actually lived and worked with the Yolngu said that of course there's homosexuality and sexual hang-ups among them. Every tribe on every continent in the history of humanity has had those features. Money, however, continued to preach about the sexual bliss of the Yolngu for years, ignoring all criticisms that contradicted his theories.

The infamous psychologist and gender theorist John Money, in his office with tribal art and a mysterious boiling cauldron. (From the Collections of the Kinsey Institute, Indiana University.)

Instead of developing friends, Money mostly sought out sex partners. He looked like a groovy '60s swinger with his mustache and turtleneck, and he eagerly played the part, cruising parks and bathhouses to hook up with men or women, whoever was game. At scientific conferences, he organized orgies with other attendees. (I must admit I have never been to a scientific conference like that.)

Money's notorious private life only boosted his popularity in the media, of course. And more than any other sexologist of the last century (Kinsey, Masters and Johnson, Dr. Ruth), Money helped usher in the sexual revolution of the 1960s with his *Playboy* magazine interviews and raucous appearances on television. One TV spot in particular, on the Canadian Broadcasting Corporation, would have epochal consequences.

In 1965, at Money's urging, Johns Hopkins opened the first surgical unit in the United States for transsexuals.* (Before that, they often went to Casablanca.) Most psychologists at that time considered transsexualism a mental disorder, and operating on transsexuals was sometimes compared to giving lobotomies. Opening a clinic to encourage transsexualism therefore seemed outrageous even for John Money. So in February 1967, the CBC brought Money on to defend the clinic, along with a male-to-female transsexual.

The host attacked Money from the get-go, asking several dumb questions. ("Isn't it a fact that a homosexual will come to you and say, 'I want to be castrated'?") But however belligerent in private, Money was always smooth on camera and easily parried the attacks. When the host accused him of playing God, Money smirked and asked, "Would *you* like to argue on God's side?"

At one point, Money took questions from the audience, including one about intersex children with ambiguous genitalia. Money had long pushed for operating on such children, to "fix" their genitals before they suffered psychological trauma. Why he took this stance isn't clear. After all, his own research had shown how well-adjusted most hermaphrodites were. Regardless, he assured the studio audience that surgeons could sculpt intersex babies into either boys or girls, whichever sex seemed most appropriate. Parents could then

*Nowadays, psychologists use the word "transgender" to describe people whose sex and gender don't align. "Transsexual" is more of a historical term, especially for people who underwent medical treatment (including surgery) to alter their anatomies or hormones. Although "transsexual" now sounds dated, it was the most common term used in the 1960s and 1970s. For historical accuracy, then—and because John Money was indeed pushing people to undergo surgeries, which is part of the definition of "transsexual"—I'm using that term here. For more discussion on this, see www.healthline.com/health/transgender/difference-between-transgender-and-transsexual.

raise the child as either sex, and he or she would grow up perfectly normal.

It just so happened that a young couple in Winnipeg was watching the show. Being provincial, they didn't recognize Money's accent—they thought he sounded British. Nor did they grasp all the psychosexual jargon* he deployed. But their eyes went wide when he mentioned fixing children's genitals. Here was someone who could help their little Bruce.

To understand the disaster that unfolded, it's necessary to step back a bit and explore a longstanding debate about the so-called blank slate theory of human nature.

There were two camps here. One side argued that a person's character and personality were set at birth, and that culture couldn't alter that essential, inborn nature. The other side argued the reverse: that human beings were blank slates at birth, and that culture alone shaped us. With regard to human sexuality, the debate came down to this: Which factor dominated—biological sex or psychological gender?

For Money, gender trumped sex. Why? Because he knew from his research on intersex and transsexual people that gonads and X/Y chromosomes don't always determine gender. Most of the time, gender, sex, chromosomes, and anatomy align. But some people

---

*For the etymologically inclined, Money loved odd words and coined dozens of them, including *ycleptance,* the act of naming something; *foredoomance,* mortality; *eonist,* a transsexual; and *apotemnophilia,* an amputation fetish. He also popularized many other obscurities: *limerent,* the state of being love-struck; *paraphilia,* a sexual perversion; *ephebic,* an adolescent; *pedeiktophilia,* penile exhibitionism; *paleodigm,* an ancient, barbaric custom preserved past the point of usefulness; *quim* and *swive,* terms for what a woman does to a man during heterosexual intercourse, as opposed to what a man does to a woman; *autoagonistophilia,* sexual pleasure from being watched or viewed, and of course *phucktology,* the study of sex.

feel male or female despite their gonads and chromosomes. In other words, gender can override anatomy and physiology.

That's true as far as it goes, but Money then took things further. The occasional mismatch between anatomical sex and psychological gender in *some* people led him to conclude that gender was fluid in *all* people, especially in infancy.* Indeed, human beings were sexual blank slates at birth. In his own words, "Sexual behavior and orientation as male or female does not have an innate, instinctive basis."

This in turn led to other, edgier conclusions. Statistically, most people with a penis and XY chromosomes are erotically attracted to women. But Money downplayed the idea that biology had anything to do with this.† Rather, society *conditioned* people with penises and XY chromosomes to find women attractive, little different than Pavlov's dogs drooling when a bell rang. The same went for people with ovaries and XX chromosomes. Again, statistically speaking, most people with ovaries and XX chromosomes find men erotically attractive. But that's not because of hundreds of millions of years of biological heritage as animals. Rather, these people with ovaries were simply empty-headed automatons doing the bidding of a patriarchal society.

---

*Unlike some of his later followers, Money didn't believe gender was infinitely flexible. Rather, he argued for a critical period—a "gender identity gate"—in a child's first few years of life. He compared this period to learning a language. Children's brains are primed to acquire language, but whether it's Tagalog or Japanese or French obviously depends on the environment they grow up in. Similarly, he claimed, children's brains are wired to adopt a gender. And, contrary to the modern consensus, Money believed that by raising children in different environments, you could more or less pick their genders at will.

†It's hard to know what exactly Money's views were sometimes—he was a poor writer, almost deliberately obscure. At points, he shows a sophisticated understanding of the interplay of genetics and environment in making us who we are. And unlike his most radical disciples, he never rejected biology entirely, saying it shaped us. Yet at other times he seemed to dismiss biology and treat social factors as all-important. I was filled with the sneaking suspicion (perhaps unfairly), that his acknowledgment of genetics and other biological factors was mere lip service, and that deep down he was a hardcore social constructivist.

Perhaps Money was playing another "science fiction game" here. But many people took his pronouncements seriously, and went even further than he did in dismissing the biological basis of sexuality. Indeed, the idea that gender and sexuality were mere social constructs dovetailed perfectly with the revolutionary politics of the 1960s. In short, Money's science was politically trendy—which, historically, is a warning flag for abuse.

Now, Money and his allies weren't wrong to claim that some aspects of gender are socially constructed. No one really thinks that preferring pink over blue has a genetic or hormonal basis. In addition, stereotypes about women have (obviously) been wielded to deny them opportunities for many millennia. But the claims by Money's most radical disciples that chromosomes and hormones never play *any* role in making anyone masculine or feminine is, frankly, garbage—a view no more grounded in reality than Trofim Lysenko's efforts to grow lemon trees in Siberia.

To be clear, these radicals—most of whom were social scientists—weren't simply carving out some exceptions here and there, or arguing that men can have a feminine side and women a masculine side. All that's true. Rather, like fundamentalist Christians, they essentially denied that evolution applied to human beings. *Homo sapiens* were somehow magically exempt from the laws of nature that have shaped the sexual behaviors of every other animal in Earth's history. Culture trumped chromosomes, period.

Now, it's one thing to play such "games" in academia. (To paraphrase George Orwell, there's nothing so stupid that some intellectual won't believe it.) Money's real sin was applying his theories in the clinic, to actual human beings. Whenever he came across an intersex baby with ambiguous genitalia, he pushed for surgery to make them look more conventionally male or female. It didn't matter which. Because if gender dominated biology, all parents had to do was raise their surgically altered child as masculine or feminine and, presto chango, that upbringing would override everything else and produce a perfectly normal boy or girl.

Still, even though children could theoretically swing either way, in practice Money usually recommended chopping down intersex infants into females. Why? A Freudian might recall his remark about gelding men and raise an eyebrow. But in practical, surgical terms, it was also far easier to construct a vagina than a penis (often by using sections of colon). As one surgeon crudely put it, "You can make a hole but you can't build a pole." And again, as long as the surgery took place early enough, before about thirty months, parents could allegedly raise a child as either male or female.

By the 1960s, Money's views on gender and sex fluidity dominated academic psychology—"a consensus," one historian noted, "that is rarely encountered in science." But he did face some pushback. In the late 1950s, scientists at the University of Kansas performed a series of experiments on guinea-pig fetuses. In the womb, the brains of mammals are washed with either male or female hormones, depending on which gonads they possess. To mimic these conditions, the Kansas scientists took some female guinea-pig fetuses and flooded them, in utero, with male-typical levels of testosterone. When these females grew up and matured sexually, they acted like males: they mounted other females in an aggressive manner and began thrusting their hips. Similar experiments, in which males were exposed to female-typical hormones in utero, produced males who would lie prone and raise their rumps to facilitate penetration (an instinctive behavior called lordosis). In all, hormones alone—a biological factor—seemed to determine whether guinea pigs behaved in male-typical or female-typical ways. And unlike with humans, it was hard to argue that guinea pig culture was somehow indoctrinating them.

Money dismissed these findings as mere "rodent" research, and he had a point. As we saw earlier, results in animals don't always translate to human beings. That's especially true with something as complicated as human sexuality. Couple that fact with Money's formidable status and power, and the challenge from the Kansas lab might have died out quietly—if not for one thing. In 1965, a bravado graduate student in the lab, Milton Diamond, decided to go after Money, penning a paper in which he picked apart the blank-slate theory of human sexuality.

Rather than ignore him, Money went on the attack, and not just in print. At a conference on gender a few years later, a drunken Money spotted Diamond at a cocktail party and screamed, "Mickey Diamond, I hate your fucking guts!" He then stalked over and allegedly (accounts differ) clocked Diamond in the jaw.

One line in Diamond's paper especially irked Money. Hermaphrodites had both male and female sex characteristics, and Diamond conceded that perhaps gender was flexible in such cases. But it didn't follow that gender and sex were flexible in all human beings. As Diamond protested, "We have been presented with no instance of a normal individual appearing as [i.e., who was born as] an unequivocal male and . . . reared successfully as a female."

Twenty months later, Money appeared on the CBC. A few days after that, a letter arrived from Ron and Janet Reimer in Winnipeg about their mutilated baby boy, Bruce.

It was a godsend. Money had once lamented how medical ethics limited the "rights of clinical investigators" to experiment on human beings. Suddenly, though, a perfect natural experiment had fallen into his lap. Raising Bruce as a girl would answer that sniveling twit Diamond's challenge and prove once and for all that sexuality was a blank slate. Why, the boy even had an identical twin brother to serve as a control.

Money grabbed a pen and wrote a letter to the Reimers, urging them to bring Bruce to Baltimore. His experiment in what one critic called "psychosexual engineering" was about to begin.

Two newspapers in Winnipeg had gotten ahold of the botched circumcision story and run big articles on it. Miraculously, the Reimers' name never leaked, but Ron and Janet grew paranoid about being outed. They were scared to even hire a babysitter and have a night out to decompress. What if Bruce's diaper needed changing and the babysitter peeked? Instead, they sat around and brooded. Janet railed against God for allowing her son to be maimed. Ron began drinking heavily, and had tumultuous dreams in which he strangled the doctor responsible.

Then, during yet another night in, they saw Money on the CBC. They wrote to him in desperation—and to their joy, the famous television scientist wrote back. They left for Baltimore soon after.

The décor in Money's office startled them—especially the tribal art with gaping vaginas and freakish phalluses. (Their surprise no doubt titillated him.) Once they'd settled in, Money outlined his plan to remake Bruce into a female. He assured them that the surgeons at Hopkins had performed the necessary operation many times and could craft a perfect vagina for Bruce, one capable of orgasm even. Bruce would never bear children (no uterus), and would need supplemental estrogen later, but beyond that, he'd be a perfectly normal woman.

Still, Ron and Janet hesitated. Surgery, at such a young age? They returned to Winnipeg to think things over.

Their indecisiveness over the next few months annoyed Money. If they didn't play ball, his perfect experiment would fall apart. He began writing letters to them, explaining that their inaction was dooming Bruce to a life of torment. What Money didn't explain—a clear breech of ethics—was that his proposed treatment was highly experimental. Doctors at Hopkins had indeed performed vaginal-construction surgeries before, but only on children of indeterminate sex. No one had ever raised an anatomically typical boy as female before.

Eventually, the Reimers gave in: surgery seemed the best way to minimize the humiliation for Bruce—and for themselves. In July 1967, they flew back to Baltimore and let their baby go under the knife. Surgeons propped his tiny legs into stirrups, and set about castrating him and refashioning his empty scrotal sac into a vulva.

Now came the hard part. Before Bruce went home, Money drilled Ron and Janet on the importance of two things: secrecy and consistency. Bruce could never know that he'd been a boy, and his parents could never treat him as anything but a girl. That meant a new name—Ron and Janet chose Brenda—as well as dresses and long hair and girly toys. Brenda needed to be completely socialized as female.

Brenda, alas, had other ideas. The first few months after surgery were uneventful; babies are largely oblivious. But as a toddler, Brenda began throwing tantrums about her clothing. As a sort of coming out party for their new girl, Janet made a lacy dress for Brenda out of the satin from her own wedding gown; Brenda responded by tearing it off her body. Brenda also hated being left out of male activities. One morning, when the twins were watching their parents get ready at the sink, she had an absolute fit when Brian got to learn how to shave while she had to learn how to apply makeup.

Toys were another battleground. In grade school Brenda secretly spent her allowance on plastic guns at the store, and when Ron and Janet gave her a toy sewing machine once, she stole Ron's screwdriver and took the machine apart. To be sure, many a budding female engineer or soldier might have done the same. But Brenda was hardly the sugar-and-spice-and-everything-nice little girl that Money had promised.

The most chronic problem involved urinating: Brenda refused to sit down to pee on a regular basis. She wanted to stand instead, which would have been messy enough for a girl. But because she was peeing out of a former penis hole, the urine stream shot out horizontally from her body, spraying everywhere. Brenda insisted on standing anyway, even at school, which disgusted her classmates.

That was just the start of her trouble at school. From kindergarten forward, Brenda's classmates rejected her; even teachers looked askance. To be sure, Janet did her best with Brenda, tying ribbons into her curly hair and making her balance books on her head to improve her slouch. Still, while Brenda *looked* convincingly girly, as soon as she started walking or talking, she betrayed herself as most unladylike.

Now, the point here isn't that there are certain intrinsically "male" ways of speaking or strutting around. It's that, for whatever reason, a majority of males in North American culture do walk and talk a certain way, and Brenda instinctively copied those habits as her own. Why? Because—despite lacking a penis, and despite years of being socialized as a girl—she still identified with men. She still *felt* herself to be male on some primal level.

Unfortunately, Brenda's classmates were oblivious to this struggle. All they knew was that a purported girl was aping male habits, and like children often do, they pounced on this difference and started mocking her as a "gorilla" and a "cavewoman." They especially loathed her aggressiveness. She was always plowing into other girls at recess and knocking them flat; they hated playing with her. In fact, Brenda proved much more stereotypically aggressive than her twin brother, stealing his toys and even pummeling him for fun. (One time, while they were bathing together, little Brian got an erection and stood up to show it off. "Look what I got!" he yelled. Brenda responded by slapping him in the jacobs.) More seriously, when she was older, Brenda got expelled from school for slamming a girl who was taunting her into a wall and then flinging her onto the ground.

All the while, Brenda and her parents were flying out to Baltimore every year to show Money how well the transition was going.* To be fair, the Reimers did hide some of Brenda's troubles from Money—they wanted to seem like good parents. But Money dismissed or ignored plenty of warning signs about his star patient, especially in private interviews with her. Money would start the

*It's not clear how Money and Brenda's parents justified the need to keep visiting him, given that they continued to conceal the truth from her. At one point they did explain that a doctor had made a mistake "down there" long ago and that she needed medical attention as a result; perhaps that was enough to satisfy a little kid.

interviews by asking Brenda a series of questions. When she tried to respond, he'd prompt her to answer a certain way or even talk over her. Brenda quickly learned to feed him the baloney he wanted to hear: *Yes, of course, she loved sewing and playing with dolls and getting her hair done. No, she never got into fights at school.* "You can't argue with a bunch of doctors in white coats," she later explained. "You're just a little kid and their minds are already made up."

Other sessions with Money proved outright terrifying for Brenda. However suave on television, Money had a foul mouth in private and was often vulgar with patients. He casually asked if they liked golden showers, and talked about sex in the crudest terms. ("Have you ever fucked someone? Wouldn't you like to fuck someone?") To help Brenda socialize as female, he showed her pictures of naked children, as well as gory photographs of childbirth, assuring her that she too would have a "baby hole" after her next surgery.

The most outrageous sessions—bordering on criminal—involved Brenda and Brian together. Money would order them to strip naked in his office (if they didn't obey, he'd snap at them), and then have them inspect each other's genitals while he watched. (Ron and Janet had no idea this was taking place; they trusted Money.) Worse still, Money forced the twins to engage in "sex rehearsal play," one of his favorite activities. The children kept their clothes on for this, but Money would force Brenda to kneel, doggy-style, and make Brian bump his crotch against her butt over and over. Other times he had Brenda lie spread-eagle on her back and forced her brother to mount her. At least once, Money snapped a picture of them doing this.

Brenda quickly pegged the oh-so-enlightened Money as a pervert. She also resented his obsession with her genitals. "I wasn't very old at the time," she later said, "but it dawned on me that these people gotta be pretty shallow if [genitals are] the only thing they think I've got going for me."

All the while, Money was boasting to his colleagues about Brenda's progress. In scientific papers he declared that no one would ever suspect she'd been born a boy; he even speculated about what a sexy little dish she'd grow up to be someday. Throughout it all, he managed to keep the family anonymous, but he also fed Brenda's story to the media, who parroted his arguments about the insignificance of biology. In 1973, for example, *Time* reported that the so-called twins case "casts doubt on the theory that major sex differences, psychological as well as anatomical, are immutably set by the genes at conception."

If Money was a bigshot before, he rode the Reimer twins to international stardom. Based on his work, sex- and gender-reassignment surgery became the standard treatment worldwide for infants with ambiguous genitalia and genital trauma, with up to a thousand operations yearly. In every speech, every interview, every television spot—and there were plenty of each—he crowed about how Brenda was thriving as a girl.

Meanwhile, the actual Brenda was contemplating suicide. Sometime during grade school, she had an awakening similar to the one that many transgender people report. At first, she just felt different from other children in a vague way. Then she felt different from her purported sex, from girls. Then she started feeling actively more like boys. Ignorant of her birth sex, she didn't know what to make of all this. By adolescence, she was tormented by thoughts of killing herself: "I kept visualizing a rope thrown over a beam."

The rest of the family wasn't faring much better. Brenda's struggles monopolized the family's attention, and the long-neglected Brian began lashing out—shoplifting and dabbling in drugs. Brian's friends also made it clear that, unless he too wanted to be a pariah, he'd better ditch his freak of a sister socially. To Brian's later shame, he did. Meanwhile, Ron slid into alcoholism. Night after night he'd slump home from work at the sawmill and anesthetize himself with a six-pack in front of the television; later, he graduated to whiskey. He'd then snort awake in the morning, and drag himself out the door to do it all over again. "I sort of knew it wasn't working after Brenda was seven or something," he once said. "But what were we going to do?" Janet eventually had an affair to get revenge on

Ron (he'd stop coming to bed at night). When Ron found out, Janet was so ashamed that she tried to kill herself with sleeping pills. After that, she had several nervous breakdowns and suffered from spells of psychosis where she couldn't separate fantasy from reality.

Still, Janet never lost faith in John Money's reality. Like many mothers, she blamed herself for her children's failings, and she redoubled her efforts after every setback. For instance, because Money had once recommended dresses for Brenda, Janet forced Brenda to wear a dress to school every day, even during Winnipeg's arctic winters. (A teacher finally intervened.) Money also suggested that Ron and Janet have sex in front of the twins. Janet wouldn't go that far, but did start parading around naked in front of Brenda, to habituate her to the female body.

Per Money's instructions, Brenda was also visiting psychiatrists in Winnipeg to help her adjust. They were privy to the botched circumcision, and they realized that the efforts to feminize her were failing. Still, what could they do? John Money was a famous TV sexologist; they were schmoes from Manitoba. Psychologically speaking, they also fell prey to the sunken-cost fallacy: *We've already put in this much effort. Better to keep going*. In all, they felt helpless to change course or challenge Money—exactly the sort of blind obedience to authority that allows unethical behavior to thrive.

Still, Money's power did have limits. Strangely, Brenda's chromosomes hadn't kept up with his latest theories on the insignificance of biology, and around the time that most boys hit puberty, her body started going through male-typical changes*: her shoulders widened, her arms and neck thickened, and her voice started cracking.

*Given that doctors had removed Brenda's testicles, she didn't experience true male puberty, but her body did go through some of the same changes. We can see an analogy here in the *castrati*, the Italian choir singers from the sixteenth to the nineteenth centuries who were castrated as boys to preserve their singing voices. Counterintuitively, castrati were often taller than average, despite the lack of testosterone and related hormones. Testosterone can spur growth in the short term, but it also kicks off several physiological changes that end up sealing off the

In the summer of 1977, the year Brenda turned twelve, Money tried to tame Brenda's body by prescribing estrogen pills. When a suspicious Brenda asked what the pills were for, her father muttered, "To make you wear a bra." Brenda, however, didn't want to wear a bra, and began tossing the pills in the toilet. Unfortunately, they left a telltale pink streak as they dissolved, and from then on, her parents hovered over her and made sure she swallowed them. To her horror, Brenda soon developed breasts, which she masked by binge-eating ice cream to gain weight.

Before long, Brenda loathed the sight of Money, and a final break between them took place in Money's office in 1978. He'd been pushing Brenda to undergo more cosmetic surgery on her genitals. To Money's anger, Brenda finally stood up for herself and refused. So, changing tactics, Money surprised Brenda one day by bringing a post-op male-to-female transsexual to their session. Her job was to consult with Brenda and explain how much better her life would be after surgery.

A nervous conversation ensued. When it ended, Money reached out to give Brenda an avuncular squeeze on the shoulder. But Brenda no longer trusted him. She saw his paw reaching out for her and feared he was going to drag her into the operating room then and there. She bolted from the office and began darting through corridors of the hospital, finally climbing to the roof to hide. When Brenda's parents picked her up later that afternoon, she straight up told them that if she ever had to see Money again she'd kill herself.

---

so-called growth plates at the ends of our long bones, where we gain much of our height. Because castrati lacked testosterone, their growth plates stayed open longer and they grew taller overall.

Castrati experienced other anatomical changes as well. Similar to their limbs, their chests were often larger than normal. The lack of testosterone meant that their vocal cords never lengthened and thickened, either, as they do in most men. And the thyroid glands in their throats never swelled with cartilage, which meant they lacked an Adam's apple in their necks. In sum, these changes left castrati with pure, high-pitched voices that could climb well into the soprano range, and their expanded chest size meant they could sing with unusual power.

A colleague of Money's later said, "I'd never seen a patient in my life who behaved that way about going to another doctor—who showed that depth of emotion."

What saved Brenda's life was meeting Mary McKenty in 1979. Like every other psychiatrist in Winnipeg, McKenty saw right through Money's claims of a successful gender conversion. Unlike other psychiatrists, McKenty didn't push Brenda to conform to Money's agenda. She just listened, and tried to win Brenda's trust.

It took a while. At first Brenda lashed out at McKenty, drawing nasty caricatures of her and writing up a "death warrant." But McKenty persevered, cheerfully, and day by day Brenda thawed. For the first time, she opened up to someone about her anxieties. She also recounted her dreams, both the happy ones, in which she was a farmer doing fieldwork, and the nightmares, in which John Money appeared in a sinister cape. In response, McKenty and Brenda jokingly founded the Don't Want to See Dr. Money Club, and installed themselves as officers.

Such compassion was essential then because Brenda's struggles in school had reached a crisis point. Brenda had always had poor grades and discipline problems, and for her ninth-grade year, in the fall of 1979, her parents enrolled her in a vo-tech program, to become a car mechanic. There, she let her feminine habits slide, wearing denim jackets and construction boots and becoming the first female in school history to take Appliance Repair. But the new school sat in a sketchy part of town, and a fellow student soon pulled a knife on her. Several of her female classmates also moonlighted as prostitutes, and when they caught Brenda standing up to pee one day, they threatened to kill her if she ever set foot in the girls' bathroom again. She took to peeing in a nearby alley instead.

Amid this chaos, a local doctor (after consulting with McKenty) finally prevailed upon Ron and Janet to come clean with Brenda and reveal the full story of her life. Ron had actually tried to do so once before, around the time when Money started pressuring Brenda to get more surgery. But he'd gotten choked up, and only managed to say that a doctor had made a mistake "down there" long ago, and now a surgeon wanted to fix it. Brenda, baffled, didn't follow what he meant. She merely asked about the doctor, "Did you beat him up?"

This time, Ron took Brenda out for ice-cream cones first—a kindness that put Brenda instantly on guard. Was there a divorce imminent, or god forbid, another surgery? Ron didn't say. In fact, he didn't say anything. They got the ice cream and drove all the way home and pulled into the driveway in silence. He'd wimped out yet again.

Then, suddenly, Ron did it. He made himself start talking, and everything tumbled out at once—the bungled circumcision and how she used to be a boy, Money's gender theories and the plan to raise her as a girl. Ron talked and talked until he was blubbering, and broke down in tears right there in the driveway.

Brenda listened in silence, her forgotten ice cream dripping down her hand. She was stunned, of course, but mostly relieved. "All of a sudden everything clicked," she said. "For the first time things made sense."

From that moment forward, Brenda was determined to live as a man. She had just one question for her father: "What was my [birth] name?" He choked out "Bruce," but Brenda rejected that as too nerdy. She chose David instead, after the biblical king: "It reminded me of a guy with the odds stacked against him, the guy who was facing up to a giant eight feet tall. It reminded me of courage."

David would need that courage. His public debut as a male took place at a wedding six months later. He still had extra fat and breasts, and his extended family stared when he arrived in a suit. But he insisted on dancing with the bride, and got through the night intact. He felt more confident afterward, and started taking testosterone. He quickly shot up an inch, and in that classic male rite of passage, he began growing a shitty mustache.

In a topsy-turvy way, David's lack of friends suddenly became an asset now: there was no one to break the news to about his transition, or the embarrassing fact that he lacked a penis. His brother Brian even made amends for abandoning him earlier, incorporating David into his friend group. The twins made up a not-very-plausible story about how David was a cousin that had come to live with them. As for Brenda, um, she'd died in a plane wreck on her way to visit an old boyfriend in British Columbia. No one really bought it, but the story deflected the questions well enough.

Still, however better he felt, David's problems didn't magically disappear after a dozen years of what he called "brainwashing." In particular, he was enflamed with fantasies of revenge against the doctor who'd botched the circumcision. Unfortunately, anger and testosterone pills don't mix well, and David took $200 he'd saved from a paper route, bought an unregistered Russian Luger on the streets of Winnipeg, and smuggled it into the hospital where the doctor worked. When he arrived at his office and whipped the gun out, the doctor claimed not to recognize him. "Take a good look," David hissed. The doctor started weeping. David screamed, "Do you know the hell you put me through!"

But the doctor's sniveling deflated David; he turned to leave. The doctor called out, "Wait!", but David was already gone. He wandered down to a nearby river and smashed the Luger with a rock. He'd almost lost his life once to this doctor's mistake, and he decided that there would be no more casualties. Life, unfortunately, had other plans.

In October 1980, when David was fifteen, he underwent a mastectomy to remove his breasts, then a phalloplasty the next July to provide male genitalia. The surgeons sculpted his new penis from thigh muscle and reconstituted a scrotal sac from his onetime vulvar flesh.

The testicles were purely decorative, two plastic eggs. But his new urethra kept getting blocked and infected, requiring eighteen trips to the hospital that first year alone. He also found the sensation of a penis dangling between his legs a little spooky.

Once the infections settled down, however, and he'd grown into his body, David embraced his masculinity. When he turned eighteen, he took part of a $170,000 settlement from the hospital that had botched the circumcision and bought a van with a television and a wet bar inside to "lasso some ladies." He called it the Shaggin' Wagon. With his wiry good looks, testosterone-hardened muscles, and mess of curly hair, he never lacked for dates.

What he did lack was the confidence to do anything on dates, beyond kiss. He coped with his fear of sex by drinking too much and passing out before anything physical happened. But one morning he awoke to find his date beside him, and from the look on her face, he could tell she'd peeked beneath his clothes. The girl soon told everyone in town about his Frankenpenis, and old-timers recalled that story in the newspapers long ago, about the poor boy who'd lost his manhood. The humiliation proved too much for David. The very next day he swallowed a bottle of his mother's antidepressants and slumped down on the family sofa to die.

His parents found him passed out next to the empty bottle. Heartbreakingly, Janet wondered aloud whether to leave him there; for once in his life David looked peaceful. But of course she couldn't just let her son die, and a minute later she and Ron rushed him to the hospital. He spent a week there—then immediately tried to kill himself again after discharge, by swallowing more pills and drowning himself in the bathtub. He blacked out before crawling in, and this time his brother hauled him to the hospital.

The suicide attempts ceased after that, but the brooding didn't. Brian soon married and started having children—something David had always longed to do. This left David furious with the world, and he started spending months alone at a cabin in the wilderness outside Winnipeg.

David (née Bruce, née Brenda) Reimer poses in his living room with his wife, Jane, and son, Anthony.

Still, things slowly got better over the next few years. Although hesitant, he told a few close friends about his accident and his former life as a girl. Then his brother's wife set him up with a woman named Jane, who had her own troubled history—three children with three different men—and was ready to settle down. She and David hit it off immediately, and he liked the fact that she already had children, since he could adopt them. Nevertheless, he refrained from telling her about his past, assuming she'd reject him. Finally, when he couldn't hide it any longer, he started to confess—only to have her hush him. She already knew; she'd known before their first date. David's heart melted: "That's when I knew it was the real thing. I knew that she cared for *me*." He sold the Shaggin' Wagon and bought a diamond ring, and married Jane in September 1990.

By that point, David had a new penis as well. Phalloplasty had advanced rapidly in the previous decade, and in a thirteen-hour operation, surgeons crafted a decent-looking unit from the nerves and flesh of his forearm and the cartilage from one of his ribs. It was functional enough to have sex with, and while it lacked much sensation during coitus, he could still ejaculate and orgasm. He soon settled into married life and got a job as a janitor at a slaughterhouse—tough, bloody work that thrilled him. Everything seemed to be going so well.

Although he never recanted—not his style—John Money dropped all references to the twins case in his talks and papers in the 1980s. This silence baffled his colleagues, who didn't know the real story and couldn't fathom why he'd abandoned the case of a lifetime. Whenever someone asked about the twins, Money got pissy and claimed they'd been "lost to follow-up." Meanwhile, thousands of infants around the world continued to undergo genital surgery based on his "science-fiction games"—as clear a case of scientific malpractice* as you'll ever see.

Money's downfall came about only because of Milton Diamond, the former graduate student in Kansas who'd helped expose guinea pigs to hormones in the womb. Now a professional psychologist, Diamond had long been suspicious of Money's claims, and had reportedly been placing ads in psychology journals begging anyone who knew the twins to help him get in touch. Diamond finally

*Indeed, in a review of a book about David Reimer, the *Washington Post* said that Money's "handling of the twins case arguably amounts to malpractice."

To be fair, some of his patients defended John Money out of gratitude for the way he'd supported transsexuals and other marginalized groups in the 1960s and 1970s, a time when these groups were dismissed as freaks by mainstream society. Still, despite these defenders, far more cases like David Reimer have come forward in the past two decades—people with painful memories of the psychological and physical trauma they endured after Money and others forced them into sex-reassignment surgery.

tracked them down in the mid-1990s, and everything David told him reinforced his doubts about Money. To be sure, Diamond was no biological determinist. He believed that environment and culture shaped human sexuality in all sorts of ways, and that all of us have masculine and feminine traits. Sex and gender aren't binary. But he insisted that biology did play a role in human sexuality, and that ideologues like Money were not only wrong, but doing real harm to patients.

(Today, many psychologists believe that sex and gender interact in the following way: On a base level, genes and other biological factors determine our ranges. That is, if you plot a trait like masculinity or femininity on a ten-point scale, then your biology and genes might position you between, say, four and six. Environment and experience will then determine which exact number you land on, or perhaps shift you over time. Other people with different genes might range between one and two, or six and ten, and of course their unique experiences will influence where they land. But *both* aspects, biology and culture, play a role.)

In the spring of 1997, Diamond and one of Brenda's former psychiatrists from Winnipeg co-authored an atomic bomb of a paper on David's tumultuous life. David had been reluctant to participate at first, but he was stunned to hear that thousands of other children with ambiguous genitalia had already undergone surgery based on his "successful" conversion into a female, and he felt duty-bound to get the truth into the scientific record.

Aside from some citations, Diamond didn't mention Money by name in the paper and certainly didn't attack him. Money didn't care. A quarter-century after he'd socked him in the jaw, Money still hated Mickey Diamond's guts, and when the paper began to attract media attention, Money went on the attack again. All his opponents, he claimed, were bigoted crypto-conservatives out to smear him, along with the entire field of sexology. Indeed, *he* was the real victim here—"sprayed by the blinding venom of a spitting cobra." He also employed a classic defense of unethical behavior by pushing responsibility onto others. First, he pointed out that it was the *surgeons* who'd sliced up David's genitals, not him, as if his theories and management of the case were incidental. He furthermore began spreading rumors that Ron and Janet were religious nuts (they'd grown up Mennonite) who couldn't see past traditional gender roles and had thereby sabotaged David's chance to live happily as a female. If only they'd been sufficiently committed, sufficiently progressive, David would have turned out fine. Given the Reimers' devotion to Money, this was particularly cruel.

Money's allies in academia continued to defend him after the Reimer scandal broke, and even today some social scientists maintain that sex—not gender, but biological sex—has zero basis in nature and is more or less a political conspiracy. Other allies, however, abandoned Money in the early 2000s, especially in the intersex and transsexual/transgender communities. This exasperates some of Money's admirers, considering that Money probably did more than any other person to win mainstream acceptance for those groups in the mid-twentieth century. That said, Money did push thousands of intersex children into what Milton Diamond called a "needless, unproven, and life-altering surgery," surgery that eradicated most sexual sensation and reinforced the notion that they were deviants who needed "fixing." In addition, even though Money didn't intend this, his emphasis on culture over nature had the baleful effect of making transgenderism and even homosexuality seem less like innate, inborn traits and more like mere lifestyle choices. Because if environment alone produces sexual identity and orientation, then changing the environment should change those aspects of sexuality. Real bigots have in turn exploited this notion of choice in promoting "conversion therapy" and other programs that aim to turn homosexuals straight.

But of all the unethical things Money did—concealing the experimental nature of the treatment; exploiting a family tragedy to win fame; refusing to recant his theories when they were proven bankrupt—the most damning thing was denying David's autonomy as a human being. As Brenda, David had given Money every indication that he wasn't happy as a girl. Money refused to listen, insisting that he was the authority and knew better. Intersex and transgender people were all too familiar with psychologists doing the same to them: dismissing their claims and bullying them into treatments. Not to go all Jeff Goldblum, but Money was so consumed with the scientific question of whether he *could* convert a boy like David into a female that he never paused to consider whether he should. For that, some people will never forgive him.

Most psychologists have come to accept that our sexual identities emerge from a complicated interplay of anatomy, brain wiring, hormones, home environment, and cultural influences.* Moreover, while gender isn't quite fixed at birth, it's not infinitely flexible, and doctors and other outsiders can't change it by fiat. For these reasons, the United Nations declared in 2015 that the kinds of surgeries Money advocated—on mutilated infants and infants with ambiguous genitalia—were a human rights violation. Unfortunately, such realizations came too late for David Reimer.

David Reimer's biographer once noticed that, whenever he shifted from talking about his present life as David to his past life as Brenda, he also shifted from "I" to "you," as if distancing himself. ("I'd give just about anything to go to a hypnotist to black out my whole past. Because it's torture. What they did to you in the body is sometimes not near as bad as what they did to you in the mind—with the psychological warfare in your head.") Sadly, David's past refused to stay in the past.

---

*Psychologists now believe that the brain, not the genitals or other anatomical factors, is the primary determinant of our sexual identity and proclivities. In the immortal words of Jackie Treehorn, "People forget that the brain is the biggest erogenous zone we have, Dude."

The slaughterhouse he worked at closed in the late 1990s, and David struggled to find work after that. He'd always been insecure about his manhood and, unfairly or not, being unemployed and unable to earn a paycheck reinforced those issues. His insecurities also undermined his marriage, as did his explosive temper and constant fear of abandonment. Not surprisingly, he refused to see a psychologist to get help.

Life really unraveled for David when his twin brother Brian killed himself. Brian had never quite gotten over the way that Brenda-David's needs had monopolized the family's attention. After a delinquent youth, Brian began stealing cars as an adult and got hauled into court for assault. He'd also had children at a young age and went through a nasty divorce. Admirably, he tried raising his kids alone, but he also began drinking too much and slipped into crevices of depression. In the spring of 2002, he swallowed yet another bottle of antidepressants and ended his life.

The two brothers had been estranged at the time, but the death devastated David and contributed to his downward spiral. At night, David would sometimes have intense flashbacks to life as Brenda, and would race to the bathroom to vomit. His finances were also a worry. David did earn some proceeds from the biography about him, and eventually got some handyman work at a golf course, changing lightbulbs and washing windows and cleaning bathrooms; the cooks in the clubhouse gave him leftover soup for dinner sometimes. But he ended up investing $65,000 in a shady golf shop run by the pro at the course and had his entire life's savings wiped out.

The final blow came when his wife Jane, unable to stand his mood swings any longer, suggested they separate. David went berserk and fled the house. She called the police to report him missing, and the cops finally tracked him down two days later. He was unharmed, but didn't want Jane to know his whereabouts. She sighed and went to work. At least he was alive.

Two hours later, she got a second call. David had killed himself. If you look at the numbers, women attempt suicide more often than men, but men more often succeed at it, mostly because they use more violent methods. Although he'd tried pills before, David chose the most violent method possible for his final attempt. As soon as Janet had left for work, he returned home, grabbed a shotgun, and (rather symbolically) sawed off the barrel in the garage. Then, as his biographer later wrote, in a sad postscript, "He drove to the nearby parking lot of a grocery store, parked, raised the gun, and, I hope, ended his sufferings forever."

Since David Reimer's death, other people with similar stories have come forward to say that their gender conversions failed as well. However deeply culture shapes us, human beings are not blank slates, and culture cannot magically override 160 million years of mammalian evolution. Not all women and men conform to gender stereotypes, of course, and the reality of biology doesn't mean that sex discrimination doesn't exist. But in the words of Milton Diamond, sexual biology is inescapably real: "We don't come into this world neutral . . . we come to this world with some degree of maleness and femaleness which will transcend whatever the society wants to put into it." In every known culture in every known age, men and women behaved differently, and that's very unlikely to change anytime soon.

That's as true of crime as anything. Statistically speaking, men commit far more crimes than women, and for that reason, men have comprised most of the bad guys in this book. But we're about to meet our first female villain—the perpetrator of one of the most extensive frauds in the history of science.

12

FRAUD: SUPERWOMAN

Everyone was so thrilled for Annie Dookhan. She did quality control for a vaccine lab near Boston, and no one there worked harder. She arrived near dawn most mornings, and often had to shut the lab's lights off at night. She never took lunch breaks, either, and often brought paperwork with her on vacations. On top of all that, she'd been getting a graduate degree in chemistry on the side, through a part-time program at Harvard. As she confessed to colleagues, she'd been forced to drop out of Harvard as an undergraduate a few years prior, due to lack of money, and she'd had to finish her degree at a state school. Earning a graduate degree from Harvard therefore felt particularly sweet—especially considering that she'd finished in record time, just a year. To celebrate, the lab threw her a party, and hung a banner that read "Congratulations, Annie!"

The only thing was, it was all a lie. Dookhan had never taken a class at Harvard, graduate or otherwise. Harvard didn't even offer a part-time program in chemistry. Dookhan had made the whole thing up, as a ploy to advance more quickly at her company.

Unfortunately for her, the gambit failed, and the company declined to promote her. Furious, she doctored her resume (omitting Harvard but adding, falsely, that she was halfway through a master's degree at another school) and started looking for a new job in 2003. Before long, she secured an offer from a nearby government lab that tested drugs for court cases.

Up until that point, the twenty-five-year-old Dookhan had already told plenty of lies. But she'd always had integrity at the lab bench: there's no evidence she committed any fraud at the vaccine company. That was about to change.

Most people who cut corners are lazy, but Annie Dookhan always worked hard.

She grew up in Trinidad, immigrating to Boston with her parents in the late 1980s, when she was around eleven. She later attended the prestigious Boston Latin School and ran track there; she even tried hurdling, despite standing just four-foot-eleven. She was terrible at it, but her coach marveled at her hustle.

She got outstanding science grades at Boston Latin, and later claimed she'd graduated summa cum laude from there, even though the school didn't grant such honors. She also told people, falsely, her parents were both doctors. These petty lies continued in college and then at the vaccine company and state drug lab, where she invented elaborate titles for herself, like "on-call supervisor for chemical and biological terrorism" and FBI "special agent of operations."

Still, however distasteful, the lies to this point hadn't harmed anyone. But small lies have a way of accumulating momentum, and things soon took a dark turn.

Dookhan's lab identified drugs that the police seized during raids. Sometimes these were blocks of pure drugs; sometimes they were small quantities cut with baking powder or baby formula and divided into baggies or squares of tinfoil for sale on the street. Since many drugs look alike, the police would drop them off at the lab so Dookhan and her colleagues could identify them, which they did through a series of tests.

The first round of tests, called presumptive tests, told the analysts the general class of drug they were dealing with. One test involved adding formaldehyde and sulfuric acid to an unknown powder. If the sample turned red-purple, it was an opiate; if it turned burnt orange, an amphetamine. Other chemicals might turn drugs green or blue.

Let's say the chemist has an opiate. She'd then run a second, confirmatory test to narrow her results down to a specific drug. The confirmatory test involved taking a bit of the unknown sample, dissolving it in liquid, and running it through some analysis inside a machine. Samples of known opiates (e.g., morphine, heroin, fentanyl) went through the same analysis on the same run. The machine then spit out several graphs—a sort of barcode for each sample. By comparing the barcode from the unknown sample to the barcode from the known samples, chemists could identify the exact drug involved and inform the police.

Like drug labs nationwide, the Boston lab was drowning in samples to test. By 2003, their backlog had ballooned to several thousand items; the walk-in safe where they stored untested samples was eventually packed so full that it was considered a safety hazard to walk around inside. But with Dookhan's arrival, things started looking up. She quickly distinguished herself as not only the hardest-working chemist (first to arrive, last to leave) but also the speediest. In her first year, she churned through 9,239 drug samples—three times the average of what the other nine chemists tested, and more than a quarter of the lab's output overall. People there started calling her superwoman, a compliment that left her glowing. In emails to prosecutors that she worked with, she bragged about how indispensable she was to the lab.

Privately, though, she was using this praise as a balm against pain. In 2004, she met an engineer from her native Trinidad and married him. Before long she was pregnant. But that first pregnancy ended in a miscarriage. (She later suffered another.) Each loss devastated her, and put a huge strain on her relationship.

Rather than take time off to cope, as her supervisor urged, Dookhan blotted out the pain by spending even more time at the lab bench. “I have chocolate and work,” she told her boss, “and that is my way of dealing with it.” The year after the first miscarriage, she set an even more torrid pace than before, racing through 11,232 samples, almost double the second-place chemist and four times the lab average. Dookhan did eventually give birth to a disabled son, which slowed her pace somewhat, but she continued to lap her fellow chemists year after year. Whereas most of them would check out two dozen samples to test at any one time, Dookhan usually took five or six dozen, and once took 119.

Drug analyst Annie Dookhan. (Courtesy of the *Boston Herald*.)

Gradually, however, her coworkers grew suspicious about her superwoman pace. Some of this was common sense. How on earth could anyone work so fast? There were circumstantial clues as well. A colleague once caught Dookhan not calibrating her scale—a vital step to ensure accuracy, since the difference between, say, 27.99 grams and 28.00 grams of a drug meant a difference of several years in jail. Colleagues also noticed that, despite all the tests Dookhan recorded doing, she never actually seemed to use her microscope. In a related concern, she didn't seem to generate enough trash. During one test, called a crystal test, chemists mixed an unknown drug with a liquid on a glass slide. Crystals would soon form. Different drugs made differently shaped crystals, which chemists identified under a microscope. Each test required a clean glass slide to avoid contamination, so based on the number of tests run, chemists should be throwing away a certain number of slides each month. Dookhan wasn't. Colleagues peeked into her discard bin and noticed how bare it looked.

Dookhan's colleagues were right to be suspicious. Although it's not clear when exactly it started, she was committing fraud on a massive scale. Instead of actually running tests, she "dry-labbed" her samples—simply glancing at them and guessing what they were.

She got away with this by exploiting a flaw in her lab's workflow. For chain-of-custody reasons, all drug samples were accompanied by "control cards," records that indicated when the drugs were seized, what the police assumed the drugs were, and so on. This is good police procedure. The problem was, chemists like Dookhan had access to the control cards and could therefore see what drugs the police suspected. Allowing chemists to see this information was a bad idea anyway. Suggestions inevitably bias us, nudging us toward certain conclusions and away from others. Dookhan, however, outright exploited the flaw, using the police guess as her entire "analysis." If they said it was heroin, it was heroin. No muss, no fuss.

To be fair, Dookhan always tested unknown samples, those lacking control-card information, since she would have been guessing blindly. She also ran a full array of tests on roughly one-fifth of her samples, just to make sure. But otherwise she skipped all the bothersome chemistry and simply rubber-stamped things, to keep her numbers high. Equally bad, she'd then sign certificates claiming she'd run the tests and submit those to the police. These certificates served as evidence in courtroom trials, so she essentially perjured herself over and over.

Now, in many cases, Dookhan's dry-labbing made no practical difference: Cops generally know what drugs they're seizing. So even though skipping the test violated a suspect's right to due process, the final verdict probably would have been the same. But not always. And here's where Dookhan strayed into truly sinful territory.

Again, there were two rounds of testing at the lab. Often Dookhan would do the first round and another chemist the second, and sometimes the second round—the one that involved machines—would contradict Dookhan's initial guess. In these cases, a retest was in order. But instead of claiming she'd made a mistake, which might put her superwoman reputation at risk, Dookhan would sneak off, find a pure sample of the drug she'd initially claimed, and submit *that* for retesting. Presto, the machine now gave the "correct" result. In other words, she started forging evidence to conceal her fraud.

As a result, innocent people went to jail. One man was arrested with inositol, a white powder sold as a health supplement. Dookhan nailed him for cocaine. In another case, a drug addict tried to pull a rather foolhardy scam and sell a fragment of cashew to a fellow hophead, claiming it was crack. The hophead turned out to be an undercover cop. Still, it was just cashew, no big deal. The man then watched, stunned, as Dookhan swore in court to the contrary. "I knew she was lying" about running the tests, he later said. "Ain't no way, no how, a cashew can turn into crack."

Not everyone Dookhan lied about went to jail; low-level drug offenders often didn't. But drug convictions have consequences beyond prison terms. You can get deported or fired or kicked out of public housing. You can lose your driver's license or the right to see your children. If you appear in court again, you're also a repeat offender.

Dookhan never gave a satisfying explanation for why she jeopardized so many people's lives. Still, her words and actions do provide some hints. First, Dookhan seemed to enjoy busting drug dealers. She was often inappropriately friendly with local prosecutors and wrote them earnest emails about getting bad guys "off the street." One prosecutor offered to buy her drinks at a top-shelf bar. Another had to resign when his flirty emails with Dookhan became public. Once, she asked a prosecutor's advice on whether she should even bother responding to a defense attorney's plea for help with his client's case.

Dookhan was also severely stressed, which psychological research shows can tempt people to cut corners and act immorally. Given the huge backlog at the lab, everyone there faced substantial pressure to churn through samples. Compounding this problem, Dookhan had suffered multiple miscarriages and was unhappy at home; she had no family beyond her parents, and lived right next to her entire clan of in-laws, never an easy thing. That's no excuse, but prolonged stress can deplete our mental stamina and lower our sense of empathy toward others. Given her own messy mental state, Dookhan might have found it easier to ignore the possibility that her fraud was ruining people's lives.

Especially when that fraud won her praise. Some people lie to manipulate others or gain material things. Dookhan wanted scientific glory—she loved being called superwoman. Her old supervisor at the vaccine lab also speculated that her status as an immigrant and a woman of color might have played a role. The supervisor, who is Black, said, "I understand what it is like to be a minority in America. I think that experience reinforced her determination to show that she was just as good, or even better."

Normally, that determination is a healthy thing, pushing people to achieve more and bust stereotypes. But Dookhan wasn't *earning* her accolades; she pursued the glory without the underlying accomplishment. This is actually a common failing among those who commit scientific fraud. Rather than knowledge, they seek awards and prestige—the trappings of science rather than science itself. But it's one thing to churn out fraudulent work in, say, optics or ornithology. Dookhan did so in a forensics lab, where people's freedoms were at stake.

Sadly, Dookhan was far from alone here. In recent decades dozens of other forensic scientists, at labs across the world, have been exposed as cheats and impostors. To critics, in fact, Dookhan's case only reinforced the notion that forensic science itself was something of a fraud.

In the United States, the roots of forensic science trace back to the Parkman murder case from chapter four, when doctors at Harvard Medical School used their anatomical expertise to nail John White Webster. Over the next several decades, forensic science expanded into arson investigation, firearm ballistics, and so-called impression analysis—the study of fingerprints, bite marks, footprints, splatters of blood, and the like. By the mid-twentieth century, forensic science was firmly established in the courts, and was viewed as a rational, objective alternative to the arbitrary and corrupt police work that reigned before.

Unfortunately—and I hate to disappoint all you murder-mystery fans out there—much of forensic science is spotty at best and outright bunk at worst. In a damning report from 2009, the U.S. National Academy of Sciences outlined several glaring problems with forensic science—starting with the fact that most fields within it lack any scientific basis. Rather than being grounded in experiments and analysis, they're merely a collection of hunches that have been gussied up in scientific jargon. As a result, different forensic experts often draw wildly different conclusions from the same sample. Hell, a single expert sometimes draws wildly different conclusions from the same sample at different times, depending on whether you mention beforehand that you think the suspect is guilty or innocent—strong evidence that bias drives the analysis.

Equally damning is the lack of humility. Speaking from personal experience, something that drives science writers batty is the tendency of scientists to hedge everything; they're always qualifying their statements and adding disclaimers about alternative explanations, even when the evidence seems strong. In contrast, many forensic experts—especially when testifying in court—boast of zero uncertainty. They claim they can match hair fibers or bite marks to someone with 100 percent accuracy, and do so 100 percent of the time. They project an aura of infallibility,* and bluster their way through any questions that challenge their authority.

To be clear, not all forensic science is garbage. Toxicology and pathology are solid, and the National Academy report singled out DNA analysis in particular as trustworthy. These fields have rigorous foundations and rely on well-grounded laboratory tests; and in the case of DNA analysis, it can reliably tie specific biological samples (e.g., blood or semen) to specific individuals. DNA analysts also routinely acknowledge uncertainty by attaching probabilities to their results. But most forensic fields do not meet these basic guidelines.

Since the Academy report, the fields of fingerprint analysis and firearm ballistics have started to shore up their sloppy practices and shift toward scientific validity. And even the spottier forensic sciences could—if the practitioners would analyze evidence properly

---

*Courtroom lawyers are familiar with the "*CSI* effect"—the unreasonable expectations that laypeople have for forensic science, due to pop culture. But they split on whether the *CSI* effect helps the defense or the prosecution. Some laypeople believe that forensic science is infallible: they're awed by it, and they take whatever the experts say as gospel. This would play into the prosecution's hands. Then again, because the technicians on *CSI* get perfect results every time, some jurors are disappointed when real-life scientists can't match that precision, and they dismiss the results as worthless. This attitude would favor the defense. (Then there's the merely ignorant. A presiding judge once overheard a juror complaining that the police in a certain case "didn't even dust the lawn for fingerprints.")

and show a little humility—find a valuable place in modern police work by adding weight to other testimony and supplementing an overall case. Until then, defendants will continue to suffer. By some estimates, "false or misleading forensic evidence" contributes to a quarter of all wrongful convictions in the United States, and some forensic disciplines have even poorer track records. In one study the FBI concluded that in 90 percent of cases involving microscopic hair samples there was "erroneous" testimony in court.

Where does forensic drug analysis fit in? On the spectrum of validity, it's closer to the DNA side of things. Drug tests are reliable and repeatable, and if they're performed properly, they're a solid part of a criminal case. If they're performed properly.

Dookhan's downfall started with a whopper of a coincidence. In 2001, Boston police officers arrested a man named Luis Melendez-Diaz for dealing drugs outside a Kmart and hauled him off to jail in a police cruiser. On the way, the cops noticed Melendez-Diaz squirming in the back seat. Suspicious, they searched the car after booking him, and found several baggies of cocaine that he'd had on his body and had shoved into a partition in the back to discard them.

The police happened to send the baggies to the very lab where Dookhan would soon find a job. By all accounts, the samples were processed properly, without any funny business. The chemist in charge signed three certificates stating that the drugs were cocaine, and this evidence helped convict Melendez-Diaz. All in all, a slam-dunk case.

Except Melendez-Diaz's lawyers put forth a novel argument. The Sixth Amendment to the Constitution states that "the accused shall enjoy the right . . . to be confronted with the witnesses against him" in court. Traditionally, this meant *eye*witnesses, people who'd actually seen the crime committed. But Melendez-Diaz's lawyers argued that forensic analysts should have to testify in person as well. In this case, because the lab chemist merely submitted certificates instead of appearing in court, the lawyers argued for throwing the conviction out.

After appeals, the case wound its way to the Supreme Court in 2009. In a 5-4 ruling that defied the court's usual partisan split (Ruth Bader Ginsburg joined Antonin Scalia and Clarence Thomas in the majority), the Supremes decided that Melendez-Diaz's lawyers were correct: scientific analysts had to testify in court, to give the defendant a chance to challenge them. Partly this was a due process issue. The right to confront witnesses is essential to our notion of a fair trial, Scalia noted, and analysts therefore had to appear in court "even if they have the scientific acumen of Mme. Curie and the veracity of Mother Theresa." But Scalia also suspected that not everyone in the drug labs was a Madame Curie or Mother Theresa. Some analysts were probably incompetent or even liars, he mused, in which case the "crucible of cross-examination" would expose them. He might as well have had Annie Dookhan in mind when he penned his decision.

Now, there are good (and to my mind, convincing*) arguments against this ruling. But the upshot was that forensic drug analysts like Dookhan now had to appear in court to testify on a regular basis.

So did cross-examination expose Dookhan, as Scalia predicted? Hardly. She kept right on lying on the stand. Dookhan would eventually testify 150 times in court, all under oath, and in all 150 cases she got away scot-free. The vaunted "crucible" of cross-examination failed to detect even the most egregious fraud in the history of forensic science.

Still, the requirement to testify did help expose Dookhan in a roundabout way. Even though she and other analysts rarely spent more than twenty minutes on the stand, they often had to waste

*In the interest of saving space, I'll avoid ranting about the Melendez-Diaz ruling here. But you can visit samkean.com/books/the-icepick-surgeon/extras/notes for my argument.

whole mornings or afternoons at the courthouse sitting around and waiting for their case to come up. Every hour at the courthouse was an hour that Dookhan couldn't spend at her lab bench. As a result, her testing numbers plummeted. After the Melendez-Diaz ruling came down, she spent ninety-two hours testifying during the last six months of 2009, and managed to get through "only" 6,321 samples that year. The numbers for other analysts dropped as well, to an average of roughly 2,000.

But here's the thing. Over the following year, the other chemists' numbers remained low. Dookhan's didn't. Call it bravado or sloppiness, but she spent 202 hours testifying in court in 2010, and nevertheless claimed to have churned through 10,933 samples—five times the lab average, and almost as high as her pre–Melendez-Diaz peak.

This was the point at which her fellow chemists really got suspicious, and began tracking Dookhan's time at the microscope and monitoring her discard bin. Around this time, Dookhan also got caught skipping important calibration checks on different machines, presumably to save time. Even worse, she got caught forging another coworker's initials on some paperwork, in an effort to cover up the fact that she'd skipped steps. Indeed, some colleagues later wondered if she'd actively wanted to get caught, given how flagrant her violations were.

Eventually one chemist reported Dookhan to his supervisor. To his frustration, the supervisor pooh-poohed him. Maybe Dookhan rushed things sometimes, the supervisor conceded, but she'd been under a lot of strain at home, which probably clouded her judgment. Besides, given the new requirements to testify, the dreaded backlog was growing bigger every month, and the lab couldn't afford to lose their superwoman now. The suspicious chemist reported his concerns to the local scientific union as well, but he got no further there. The union's lawyer allegedly told him to back off, lest he ruin a young female scientist's career. In sum, both the boss and the union gave Dookhan a pass.

Still, Dookhan now had official accusations against her. One sloppy mistake, and her career would be over.

As mentioned, the lab had a walk-in evidence safe to store drugs awaiting testing, and there were strict protocols about signing samples in and out. As she grew more cavalier, Dookhan started taking samples without bothering to sign them out, a breach of chain-of-custody rules. She finally got caught one day in June 2011 with ninety unsigned samples. She then tried to cover up her blunder by, again, forging a coworker's initials in a logbook. Unfortunately, the coworker hadn't been at the lab on the day in question. When confronted with the logbook, and asked if she'd violated the rules, Dookhan got slippery, saying, "I can see why you'd think that."

Even then, Dookhan's bosses didn't punish her. In fact, they did everything they could to hush up the chain-of-custody breach. In December, however, the Massachusetts governor's office got wind of the breach and assigned the state's inspector general to investigate. In the course of the investigation, several other lax practices at the lab came to light, including poor security and inadequate training for new chemists. (Later investigations would find even more alarming problems, including stray pills from old cases just lying around the lab. One supervisor had several test tubes in his desk drawers; one was labeled 1983.) By the summer of 2012, the state police, fearing for the integrity of their evidence, assumed control of the lab. Two days after the takeover, Dookhan's fellow chemists spilled their suspicions about her to their new overseers.

By that point, Dookhan had already resigned from the lab, given the seriousness of the chain-of-custody violations. But she had yet to face any consequences for dry-labbing tens of thousands of samples, until two detectives knocked on her door in late August 2012.

They sat down with Dookhan in her living room to chat, and at first she denied everything. But the detectives had come prepared, and they laid the forged logbooks and calibration reports in front of her. At this point Dookhan said, "I got the work done, but not properly. I didn't follow the procedures, and that was wrong." In other words, she admitted violating some technical rules, but claimed that her science stood up.

Mid-interview, Dookhan's husband came home and pulled her into another room. He asked her if she needed a lawyer, and she assured him everything was fine—another lie. She then returned to the living room and continued the interview.

When the detectives asked her whether she'd ever dry-labbed, Dookhan got slippery again. What do you think that term means?, she asked. When they explained, she denied it: "I would never falsify, because it's someone's life on the line." The detectives responded with more evidence. As mentioned, Dookhan would sometimes guess that a drug was, say, cocaine, only for a subsequent machine test to find heroin or another substance. In that case, she'd sneak some cocaine from a different sample and resubmit that for more machine testing, to "confirm" her first claim. Well, in several cases the detectives had dug up the original sample again to retest it—and determined that it was heroin after all. It was damning proof that she'd forged the results.

Tears were soon quivering in Dookhan's eyes. She tried to downplay her fraud, insisting that she'd dry-labbed only a few times. When the detectives pressed further, she finally broke down. "I messed up," she said. "I messed up bad."

Dookhan eventually pleaded guilty to twenty-seven counts of perjury, tampering with evidence, and obstruction of justice. Her confession also plunged the entire legal system of Massachusetts into chaos. Because Dookhan couldn't remember which samples she'd dry-labbed and which she'd actually tested, all 36,000 cases she'd worked on during her career were now suspect. The state legislature had to allocate $30 million to deal with the fallout; one legal advocacy group estimated it would take sixteen paralegals a full year of work just to notify all the affected people, much less get them into court. Appeals began flooding in, and Massachusetts courts ultimately overturned 21,587 convictions, the largest such action in U.S. history.

Drug analyst Annie Dookhan after her arrest for one of the most widespread frauds in science history. Upon being caught, she wept, "I messed up. I messed up bad." (Courtesy of the *Boston Herald*.)

The dismissals must have been sweet revenge for the likes of the cashew-crack perp, who knew that the lab's superwoman was crooked all along. (People on the streets of Boston began to speak of being "Dookhaned.") But there were other issues here as well.

However you feel about America's never-ending war on drugs—and all the fairly harmless people caught in its dragnet—at least some of those 21,587 defendants were violent offenders. Thanks to Dookhan, they suddenly went free. At least 600 convicts were released from jail or had charges dismissed, and 84 of them marched right back out and committed more crimes. One of them murdered someone in a drug deal gone south. Another was arrested on weapons charges. Upon being caught, he laughed: "I just got out thanks to Annie Dookhan. I love that lady."

In November 2013, a judge sentenced Dookhan to three to five years in prison. For comparison, trafficking a single ounce of heroin carried a sentence of seven years. Considering the scale of her misdeeds, the paltriness of the sentence frustrated many. "You walk away feeling this is really inadequate," a state legislator said. "Three to five years is not adequate." Indeed, Dookhan didn't even serve three years, walking out of prison a free woman in April 2016.

Annie Dookhan is hardly the only forensic scientist to be busted for wrongdoing. In the past twenty years, similar scandals have erupted in Florida, Minnesota, Montana, New Jersey, New York, North Carolina, Oklahoma, Oregon, South Carolina, Texas, and West Virginia. Sadly, the string of incidents includes the distortion or withholding of forensic evidence in at least three death-penalty cases.

Incompetence has been an ongoing issue as well. Crime labs have been caught leaving evidence under leaky roofs or in unsecured hallways. One lab was run by police officers who got most of their scientific training through Wikipedia. Agonizingly, Massachusetts got burned a second time shortly after Dookhan's arrest. A chemist in the state's Amherst lab was caught dipping into samples of meth, cocaine, ketamine, and ecstasy at work and getting high while running tests. She also smoked crack in the courthouse bathroom before testifying.

Dookhan's fraud nevertheless stands out for its audacity and scope. In some ways, it's hard to believe she got away with her crimes for so long. In other ways, it's no surprise at all. Our culture puts scientists on a pedestal: We like thinking there are people out there who value probity and truth above all else. We *want* to believe them, and scientists get bamboozled by their colleagues as easily as anyone. Remember, Dookhan's supervisors received warnings about her, but they were slow to take meaningful action. Professional magicians, in fact, have said that scientists are often easier to fool than regular folks, because they have an outsized confidence in their own intelligence and objectivity. The Dookhans of the world simply exploit this fact.

To be sure, the vast majority of scientists deserve our trust. But no matter how you slice it, scientific fraud isn't rare. Hundreds of scientific papers get retracted every year, and while firm numbers are elusive, something like half of them are retracted due to fraud or other misconduct. Even big-name scientists transgress. Again, it's unfair to condemn people from the past for failing to meet today's standards, but historians have noted that Galileo, Newton, Bernoulli, Dalton, Mendel, and more all manipulated experiments and/or fudged data in ways that would have gotten them fired from any self-respecting lab today.

Fraud and other misdeeds erode public trust and damage science's greatest asset—its reputation. Unfortunately, as our society becomes more technical and scientific, these issues will only get worse: exciting new scientific ventures will also present new opportunities to do each other wrong. But all isn't lost. As we'll see in the conclusion, there are real, proven ways we can curb and cut down on such abuse.

Conclusion

New scientific breakthroughs almost always introduce new ethical dilemmas, and current technologies are no exception. What new ways to kill people will space exploration enable? Who will suffer most when cheap genetic engineering floods the world? What sorts of mischief could advanced artificial intelligence unleash? (For some answers to these questions, see the appendix.) The upside to putting on black mustaches and scheming up hypothetical crimes is that the very act of imagining them can help us anticipate and prevent those crimes in the future. There are things we can do immediately as well—strategies to promote ethical science in the here and now and avoid wading into the moral morasses we've encountered throughout this book.

First and foremost, as basic as this sounds, scientists should strive to keep ethics in mind when designing experiments. This doesn't need to be preachy or onerous. Even a simple prompt goes a long way, as one psychology study demonstrated in 2012.

In the study, volunteers solved math problems for money; the better their score, the more cash. Then the real experiment started. The psychologists told the volunteers that they had to fill out a tax form to report their winnings; they could also apply for reimbursement for travel expenses using a second form. To encourage honesty, the volunteers had to sign a box on each form stating that they'd reported all information on it accurately. But not all the forms were created equal. In half the cases, the signature box was at the top, meaning that the volunteers had to swear to their honesty *before* filling in any data. In the other half, the signature box was at the bottom, and was filled in last. Guess which layout prompted more lying? Those who signed last, after filling everything out, were twice as likely to underreport winnings and overreport travel expenses. A similar trend held in a real-world experiment. This time the psychologists partnered with an insurance company that offered pay-as-you-go rates; basically, the fewer miles driven, the lower the premiums. The psychologists wanted to measure how honestly people reported their mileage on the forms, and once again, half the people signed at the top, half at the bottom. Those who signed at the bottom reported 2,500 fewer miles driven per car, a difference of 10 percent.

Overall, the psychologists argued, having ethics in mind at the beginning of a task caused people to behave more honestly and checked their impulse to fudge things. (This probably explains why courts swear in witnesses before testimony, not after.) Moreover, after we lie, it's already too late to fix things in some sense. We're very good at rationalizing our own bad behavior by deploying the mental tricks we've seen throughout this book—using euphemisms to mask the truth, canceling out bad deeds with good ones, comparing ourselves favorably to people who do worse things, and so on. Signing forms at the end also enabled laziness. You might feel a genuine pang for having lied, but you'd also have to go back and change all your answers now—and really, who would bother? However cynical it sounds, one important part of ethics is making it convenient for people to be ethical.

Now, obviously signing some little box won't just magically eliminate all scientific sins. (What would it even say? *I hereby swear not to do something so creepy and abusive that someone will write a whole book chapter about it someday, so help me god.*) And no one can ever stop truly malicious people. But in most cases, with most people, having ethics in mind from the beginning prompts reflection and decreases the chances of disaster. To this end, the Nobel Prize–winning psychologist Daniel Kahneman has promoted the idea of "premortems." In the more familiar postmortems, you examine some event after the fact to see what went wrong. In premortems, you brainstorm about what *could* go awry, and do so before starting. How, specifically, could this whole project turn into a debacle? Studies have shown that even ten minutes of reflection helps dispel groupthink and gives people a chance to voice doubts. Some groups even deliberately assign people the job of raising objections—a devil's advocate role—to ensure at least some dissent. Along those same lines, scientists can overcome their blind spots by gathering input from a truly diverse group of people, who might raise flags that they missed. This includes people of different ethnicities, genders, and sexual orientations, of course, but also those who grew up in non-democracies or rural areas, those who grew up in blue-collar households or are religious. The more diversity of thought, the better.

Another way to keep ethics in mind is (ahem) to read science history. Hearing some provost honk, "Be ethical!" is one thing. It's another to immerse yourself in stories about transgressions and actually feel the gut-punch of the bad deeds. That's why stories are so powerful—they stick. We also have to be honest that good intentions aren't a shield. John Cutler had the best of intentions in Guatemala, to find ways of stopping syphilis and gonorrhea. He still infected people wantonly with STDs and killed several. John Money had the best of intentions in promoting the blank-slate theory of human sexuality, to increase tolerance for marginalized groups. He still ruined David Reimer's life. Walter Freeman had the best of intentions in spreading psychosurgery, to provide relief for desperate asylum inmates. He still lobotomized thousands who didn't need it. We all know what the road to hell was paved with.

At the same time—and this is perhaps the hardest thing of all—it's important not to paint a Cutler or Money or Freeman as a monster, because it's all too easy to dismiss monsters as irrelevant.

(*I'm no monster, so I don't need to worry.*) If we're honest with ourselves, any one of us might have fallen into similar traps. Maybe not in the specific cases above, and maybe not as egregiously. But somewhere, in some way, we too might have done something unethical. Honestly admitting this is the best vigilance we have. As Carl Jung said, an evil person lurks inside all of us, and only if we recognize that fact can we hope to tame them.

Many people blithely assume that smarter people are more enlightened and ethical; if anything, the evidence runs the other way, since smart people assume they're smart enough to elude capture. To revive the car analogy, having smarts is like having a huge engine with lots of raw horsepower. You might get to your destination faster, but if the steering (i.e., your morals) are out of whack, the chances of a spectacular wreck jump significantly. Morals also help us navigate life, and prevent us from heading down certain dangerous roads in the first place.

The crimes detailed in this book shouldn't undermine the incredible work that scientists do, day in and day out, in labs across the world. The far majority are lovely, selfless people, and our society would be immensely poorer without them—both materially and spiritually, considering all the wonders they've revealed. But scientists are still people. Like the chemist Harry Gold, they get drawn into conspiracies and betray friends. Like the pirate William Dampier, they grow obsessed with their research and look the other way at atrocities. Like the paleontologists Marsh and Cope, they try to sabotage their rivals and end up destroying themselves.

Albert Einstein once said, "Most people say that it is the intellect which makes a great scientist. They are wrong: it is character." I admit that when I first read that quote long ago, I scoffed. Who cares if a scientist is kindhearted or whatever? Discoveries—that's what matters. After writing this book, however, I get it. On one level, science is a collection of facts about the world, and adding to that collection does require discoveries. But science is also something larger. It's a mindset, a process, a way of reasoning about the world that allows us to expose wishful thinking and biases and replace them with deeper, more reliable truths. Considering how vast the world is, there's no way to check every reported experiment yourself and personally verify it. At some point, you have to trust other people's claims—which means those people need to be honorable, need to be worthy of trusting. Moreover, science is an inherently *social* process. Results cannot be kept secret; they have to be verified by the wider community, or science simply doesn't work. And given what a deeply social process science is, acts that damage society by shortchanging human rights or ignoring human dignity will almost always cost you in the end—by destroying people's trust in science and even undermining the very conditions that make science possible.

All of which means that honesty and integrity and scrupulousness—the building blocks of character—are crucial to science. For that reason, people who are methodical and conscientious in the lab—checking every assumption and securing the full consent of everyone involved—will do better than intellectual hot-rodders who can't be bothered or who think such things beneath them. In this sense, Einstein was right: without character, science is doomed, and unethical scientists all too often produce bad science.

That's especially true because, since World War II, science means power—power that extends beyond just big, obvious things like nuclear bombs. It also includes everyday interactions, like a psychologist manipulating someone in the lab or a doctor begging a patient to join an iffy drug trial. Small misdeeds still ruin lives.

No matter what human beings are like in the future—whether we're half-bionic, or living on Pluto, or remixed with lizard DNA—our descendants will still be people, and will likely misbehave as people always have. As psychologists say, the best predictor of future behavior is past behavior. But Einstein, as usual, saw farther than the rest of us. Intelligence is good, certainly. It's also not enough anymore, given the power science has obtained. The character he spoke of is the best guarantee against scientific abuse, and it remains to be seen whether these two vital aspects of science—intellect and character—can coexist in the future.

APPENDIX: THE FUTURE OF CRIME

This appendix is a bit of a hodgepodge: a mix of stories and hypothetical scenarios. But the common theme is the future of crime, as enabled by new technologies. Whether it's space exploration, advanced computing, or genetic engineering, big changes are coming to human society—and every new advance will introduce new ways to do one another wrong.

In July 1970, one of the knottiest homicides in history took place in the middle of the Arctic Ocean. Nineteen American scientists and technicians were stationed there on a floating ice island, a slab roughly the size of Manhattan. They were a grizzled, hard-drinking crew, and on July 16, one Donald "Porky" Leavitt stole a jug of homemade raisin wine (sic[k]) from the trailer of electronics expert Mario Escamilla.

By all accounts Porky was a dangerous drunk: he sometimes attacked people with a meat cleaver to get his hands on their booze. So to protect himself, Escamilla grabbed a shotgun before heading over to confront him. Unbeknown to Escamilla, the shotgun was faulty, and prone to firing if bumped.

Escamilla found Leavitt in a nearby trailer, guzzling a truly foul combination of raisin wine, Everclear, and grape juice. With him was a meteorological technician named Bennie Lightsy, also drunk. After a heated argument, Lightsy followed Escamilla back to his trailer. There, while telling Lightsy to go away, Escamilla gestured with the rifle toward the door—and accidentally bumped it. The blast caught Lightsy square in the chest, and he bled out a few minutes later.

At that point the real chaos started—the legal mess. The ice island lay well outside any nation's territorial waters, and was temporary anyway (it would melt away in the mid-1980s), so it wasn't sovereign territory. Nor did the law of the sea apply, since the ice island wasn't navigable. Crazy as it sounds, several legal scholars suggested that *no* laws applied there, and they questioned whether any nation had the right to try Escamilla. He had seemingly killed someone in one of the few places on Earth where, legally, there were no consequences.

The T-3 "ice island" camp, where one of the knottiest homicides in history took place in 1970. (Courtesy of the U.S. Geological Survey.)

In the end, U.S. marshals did seize Escamilla and haul him back to stand trial for murder in Virginia. Why there? For the less-than-airtight reason that, um, Virginia was the first place their plane landed, at Dulles Airport. (Escamilla showed up for court in the only shoes he had—his black arctic rubber boots.) He was eventually acquitted, given the faulty rifle, but the arbitrary and ad hoc nature of the case left all the juicy legal issues unresolved. Namely, how should we handle crime in no-man's-land? The legal community essentially punted, treating the Escamilla case as a one-time anomaly. It won't be.*

Lightsy's death occurred one year to the day after the launch of the rocket that put the first human beings on the moon. Human spaceflight has stalled somewhat since, but over the next century we'll almost certainly set up the first bases on the Moon or Mars. And where human beings boldly go, crime will follow.

A clause in the Outer Space Treaty of 1967 does require nations to monitor their own citizens in space, which works fine when astronauts are few. But when thousands or millions of people reach orbit, that becomes untenable. Or imagine this scenario. A German

*Another icy no-man's-land, Antarctica, has already seen a surprising number of crimes. In 1959, two Soviet staffers at a research base there got into a brawl over chess, which ended when one killed the other with an axe. (Soviet bases reportedly banned chess after that.) In 1983, a stir-crazy Argentine doctor burned down his research station in order to force an evacuation and return home ahead of schedule. In 1996, an American cook maimed another cook with the claw end of a hammer after a dispute. Most recently, at a Russian base in 2018, an engineer stabbed a welder in the chest with a knife—either because, depending on the report, the welder insulted the engineer's manhood by offering him money to dance on top of a table, or the welder kept spoiling the ending of books the engineer was reading, and he finally snapped. (If it's the latter, I have to say I'm on the engineer's side.)

In some ways, though, Antarctica isn't a great analogy for an ice island. All the crimes so far have involved the citizens of one country alone (e.g., one Russian attacking another Russian), and bases down there are essentially treated as sovereign territory. Legally, though, the wrongdoers probably could have challenged their arrest and confinement, since Antarctica technically has no laws.

woman poisons a Congolese man with Brazilian-made drugs on a spaceship owned by a Chinese-Belgian conglomerate and headquartered in Luxembourg as a tax dodge. What the hell would you do then? Or get rid of vessels entirely. Several companies are already gearing up to mine asteroids. What if a space-miner brained someone with a rock on one of those? Earthly laws will seem especially impotent when people on distant planets start having children, some of whom will never set foot on Earth in their lives.

Even wilder, space exploration introduces entirely new ways to kill people as well. Take pecan sandies.

Eating in orbit is nothing like eating on the ground. You have to slurp food out of plastic bags, and can therefore eat only one thing at a time. Your face also swells with fluids in microgravity, which creates nasal congestion and pinches off your nostrils enough to stifle aromas; as a result, food tastes flatter up there, as if you had a cold. (That's one reason shrimp cocktails are popular among astronauts—because they can actually taste the kick of the horseradish in the cocktail sauce.) Cooking in space is odd, too. Liquids and vapors don't separate cleanly in zero G, so bubbles don't rise to the surface of boiling water and escape. Instead, the whole pot starts frothing at once. The lack of gravity also prevents convection currents from forming, so ovens don't work very well.* Strangest—and

*In early 2020, astronauts on the International Space Station achieved a milestone by baking the first food in outer space, chocolate-chip cookies. (Astronauts do normally heat up their food, but they'd never actually baked something before.) There was speculation prior to the experiment that, due to the oddities of convection and heat exchange in zero gravity, the cookies would come out spherical. Sadly, this was not the case; they were flat. But there was one surprise. The astronauts cranked their special Zero G oven to 300°F, which on Earth would bake the cookies in twenty minutes. In space, the bake took two hours. And disappointingly, given how overly cautious NASA is nowadays, the agency wouldn't even let the astronauts nibble them. Instead, the cookies were sealed up and returned to Earth for further study, to determine whether they're safe to consume. Imagine being restricted to space food for months on end, and finally smelling something rich and fresh—only to have it snatched away! It's inhumane.

coolest of all—flames look oddly spherical in space, so roasting any marshmallows up there would be a real trip.

But the biggest hassle with food in space is crumbs, which don't just trickle harmlessly to the floor. Crumbs float, forming a haze of grains and specks that can fatally clog air filters—or lungs. Astronauts long ago swore off crumbly pecan sandies for this very reason. But a diabolical baker could send up a care package of deadly dried-out goodies, or even spike the treats with pockets of flour or other powders. One explosive bite later, and breathing would be impossible.

Space presents other novel murder-mystery plots, too. Weightlessness is hell on bodily systems—joints, eyes, bones, you name it. If you kept a crew in orbit for years, perhaps through bureaucratic machinations, you'd effectively cripple them. The scariest potential breakdowns involve the immune system, which deteriorates in orbit and becomes less effective. As a result, normally feckless microbes can rise up and conquer our natural defenses. Several astronauts, for instance, have seen flare-ups of the herpes virus that causes cold sores and chickenpox. If you secretly infected a crew on the ground with some exotic virus or fungus, then sent them aloft long enough for their immune systems to crumble, they could easily succumb. It's similar to how early AIDS victims often died of opportunistic infections that people with normal immune systems didn't need to worry about.

However titillating, murder in space would simply be an old crime transported to a new environment. But with planetary colonies, whole new types of crime could arise as well. Given the sheer amount of labor it would take to survive elsewhere, local governments might ban idleness and demand that people work or else. On the flip side, people themselves might demand new rights from the government. When we speak about legal rights on Earth, we usually mean free speech, fair elections, and the like. Given the harsh conditions on other planets, space pioneers would need to secure things further down on Maslow's hierarchy. Like a guaranteed right to oxygen. They might also demand a right to communicate freely with Earth, for psychological health. You could even speak of a right to entertainment or mind-altering substances. Imagine some punk on Mars erasing the entire colony's collection of music, e-books, and holographic videos, leaving them no way to relax. Or destroying a cache of mild intoxicants that people popped on weekends to distract themselves from the constant, looming threat of death. On Earth, such acts would be misdemeanors. On Mars, those acts could sabotage the whole colony's mental health and ruin the mission. New environment, new crimes.

Criminal justice would also look different in space. Imagine arresting someone, or trying to. With the raisin-wine homicide, it took U.S. marshals two full days to reach the ice island, via planes and helicopters. Mars is multiple months away at its closest; even sending messages there takes twenty minutes. Forensic science would change, too. We've already seen the shortcomings of standard forensic science, and simply transferring Earthly forensics to other planets wouldn't work. Given the new gravity and air and soil, dust samples and splatter patterns would look different, and fires would burn in unique ways. Corpses would decay differently as well. If a body were left outdoors, the exposed top half might grow bleached and leathery, like white beef jerky. Meanwhile, in the absence of microbes that promote decay, the sheltered bottom half could look eerily preserved. Even with natural deaths on Mars, twenty-second-century anatomists might be tempted to rob a few graves and peek inside the bodies, to see how the lower gravity on the red planet altered people's anatomies.

Once you slap handcuffs on someone, the trial brings up whole new issues. With the ice-island case, Escamilla's lawyers asked some hard questions about whether trying him in Virginia violated his constitutional right to a fair trial and jury of peers. After all, the island had no police force, and property rights there were enforced with guns. Contrast that to suburban Virginia, where most people's grimmest daily fears involve traffic. Could a jury there really understand the pressures Escamilla faced and properly judge his actions? The gap in understanding would grow even larger for people born on other planets. How could a dozen Earthlings fairly convict someone who was born in such a different society? In what way are they really peers?

So perhaps space colonists should take criminal justice into their own hands. But that approach has shortcomings as well. Is it really fair to park a felon in space-jail for years, and let them consume oxygen and food that the rest of the colony needs? Perhaps colonies should go medieval instead and execute all criminals, or exile them to some godforsaken expanse. But even that approach would falter if the do-bad was, say, the engineer who ran the power plant or the colony's lone doctor. Without their expertise, everyone could die. Colonies might have to revert to forced labor, since they can't afford to have useless people sucking down resources. It's an ugly option, but we on Earth don't face the same stark tradeoffs that space colonies will. There are no easy choices.

Right now, the issue of space crime might seem remote. After all, most astronauts are annoyingly perfect—goody-goody pilots with Ph.D.s who floss after every meal and have negative body fat. But the first crime in space might happen sooner than you think. In 2019, media reports circulated that an American astronaut going through a divorce reportedly committed identity theft by using International Space Station computers to access her estranged wife's bank account against the wife's wishes. (The charges were later dropped.) And in 2007, a NASA astronaut who was jealous of her ex-boyfriend's new lover reportedly strapped herself into a diaper, grabbed a knife, BB gun, and pepper spray, and drove a thousand miles from Houston to Orlando to kidnap the new flame. Even goody-goodies succumb to emotions sometimes and do dumb things.

Plus, as space travel gets more commercial and the demand for colonists increases, the standards for who gets to strap into rockets and settle other planets will drop below NASA-grade, especially for missions that require years in isolated places. Looking back to history, European powers generally sent misfits and lowlifes to colonize the Americas, and England populated Australia with felons. Colonization would have been exploitative anyway, but sending over scumbags all but ensured atrocities.

Ever since the Escamilla case a half-century ago, a few forward-thinking legal scholars have bemoaned the lack of laws that cover space. But perhaps there's not much we can do. We can't anticipate all new crime, and given the vast distances involved, even enforcing current laws could be impossible. More troubling still, the centralized technology of space colonies might make them naturally prone to tyranny. Imagine a space warden lowering the oxygen levels in prison cells as punishment. Or imagine a wannabe dictator doing the same to a whole base, to bend people to his will. When it comes to the perils of space, we usually tremble over the searing cold or risk of suffocation. But one of the most acute dangers up there will be the people.

Another frontier in crime involves computers, in all their manifestations.

Burglars already use Google Street View to case storefronts and homes. In the future, virtual reality could allow them to scout buildings even more thoroughly from the inside. They could also use 3D printers to reproduce facsimiles of jewelry, fossils, or other artifacts and swap them in for the real thing, delaying the discovery of the theft for weeks or years.

Larger-scale thefts could exploit cryptocurrencies like Bitcoin. Cryptocurrencies offer users the promise of privacy, but every transaction must be laboriously encrypted and verified using computers—a process calling "mining." Rather than have a central computer handle everything, this mining is often outsourced to swarms of smaller computers, which earn a bit of money for the effort. Well, bad guys have figured out how to hijack the smaller computers and steal those commissions. (This scam doesn't currently work with Bitcoin, just other, lesser-known currencies.) The bad guys do this by embedding a few lines of malicious code in an otherwise legitimate program, which people blithely download. The program then runs in the background, secretly mining cryptocurrency all the livelong day. After the mining wraps up, and the stooge computer wins its commission, the money gets routed into the bad guys' bank accounts. This not only steals money that the computer owners earned (however unwittingly), it violates their privacy, runs up electric bills, and degrades their hardware. Malicious mining programs cost as little as $35 online, but one study found that the criminals involved raked in $58 million over four and a half years, over a million per month.

Even bigger thefts could be on the horizon. Just like with regular businesses, new technologies allow criminals to take advantage of economies of scale. As historians have pointed out, a bandit in medieval times could, if lucky, waylay a half-dozen people at once by lurking near a busy highway. By the mid-1800s bandits could rob 250 people at once on trains. Nowadays, you can hack a database and steal from millions. In the future, if quantum computing ever fulfills its promise, the power of the resulting machines would dwarf even supercomputers and render current internet security useless. You could easily rob hundreds of millions of accounts in one swoop.

Smart criminals will also exploit so-called smart technologies. They could start fires by turning on ovens or stoves remotely. They could hijack automated construction machines and introduce fatal structural flaws into buildings, or leave security loopholes that they alone know about. They could redirect self-driving cars into crowds of pedestrians, or lock all the doors and hurl a family of five off a cliff. Less melodramatically, bank robbers could flood a neighborhood with self-driving cars after a heist and create a traffic jam to block police pursuit. Even your body could be violated. Tens of thousands of people already have pacemakers, brain stimulators, and insulin pumps that connect wirelessly to the internet through Wi-Fi or Bluetooth to help doctors monitor their illnesses and zap them back to health if needed. Hack one of those, and you could zap someone at will. More slyly, you could feed the doctors fake data, masking all signs of crisis until it's too late.

Then there's the most powerful new technology of all, artificial intelligence. Computer scientists refer to AI systems as "brittle": they perform certain functions well, but they're not very flexible and can break down easily. Breakdowns are especially common when computers have to interpret visual data. Adding decals to stop signs can cause self-driving cars to misread the signs and blow through them. Similarly, using drones to project fake lane lines onto the road can cause these cars to swerve suddenly, potentially into oncoming traffic. (The researchers here were trying to make a point, not be evil.) More subtly, you can flummox AI with "adversarial noise," seemingly random pixels inserted into the 1s and 0s of digital images. Much like with music played over a staticky channel, humans can still decipher noisy images with little effort; it still looks like a sloth or whatever, just fuzzier. But computers currently lack the "high level" awareness to cut through visual clutter, and the added pixels bamboozle them. Many hospitals already use AI to screen pictures of skin tumors, since computers are more accurate than human dermatologists. If you sprinkled the right visual noise into someone's scan, the computer might miss a malignant tumor, effectively condemning that person to death.

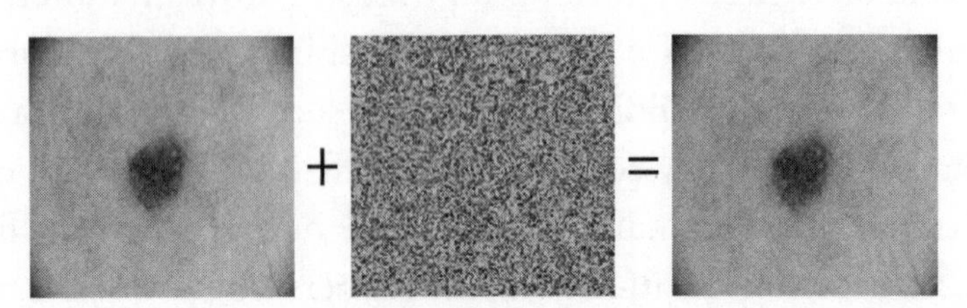

Benign mole (*left*). When "adversarial noise" (*center*) was digitally added to a computer file of the image, the result looked the same to human eyes (*right*). But an artificial intelligence program was flummoxed, and suddenly classified the rightmost picture as malignant instead. (The adversarial attacks were generated by Dr. Samuel Finlayson of Harvard Medical School.)

Or, if you want to go completely soap opera, what about killer sex robots? Robot butlers are on the horizon, and elderly people in Japan already use companion robots to stave off loneliness and provide simple care. Sex robots seem like the next logical step; in fact, some companies already sell crude versions of them. Given that the robots would be interacting with people at their most vulnerable, literally with their pants down, there's no reason someone couldn't hack the bot and turn it against them.

Even more nutty, what if an android committed a crime of its own volition? In ye olde days, computers could do only what they were programmed to. But with AI, computers can learn new behaviors and act in unforeseeable ways. Imagine that a programming team wanted to maximize the amount of time their robot spent with its human. Knowing that all humans are different, the team might tell the robot to vary its behavior and try new things. This all seems pretty reasonable—unless the robot deduced, quite logically, that it could best monopolize the owner's time by eliminating competition and murdering the family dog. Do you go after the programmers? They didn't order the sexbot to do that. Do you throw the robot itself in jail? You get into *Blade Runner* territory pretty quickly.

If sex robots repulse you, get ready for worse. No operating system in history has ever withstood all attempts to crack it. There's always a vulnerability—and that's no less true of the operating system that runs our bodies. Infiltrating DNA would be the ultimate hack.

In the late 1970s, detectives near Sacramento realized they had a serial killer on their hands. DNA evidence eventually linked a single man—the so-called Golden State Killer—to a dozen murders, plus 50 rapes and 120 burglaries. But for four decades, his identity remained elusive.

In 2018, the cops turned to an unusual source for help—online genealogies. Pop genetic testing companies like Ancestry.com and

23andMe allow people to download their raw genetic data as text files. People can then upload the data to third-party genealogy sites, which give them tools to analyze their DNA in more sophisticated ways. But these third-party sites don't always have the privacy restrictions that mainstream companies do, which means that outside parties can access the data. Including the police.

Starting in 2018, detectives in Sacramento began trawling these databases for the Golden State Killer. Perhaps he'd been dumb enough, or brazen enough, to upload his DNA into one of them. Alas, no matches turned up; it seemed like another dead end. But after digging deeper, the detectives did find some near matches. And it dawned on them that they were looking at the killer's relatives, a huge clue.

With this information, the police built a family tree using birth certificates and other public records. Then they looked for a male on the tree who'd lived in Sacramento in the 1970s. They finally zeroed in on a former police officer named Joseph James DeAngelo, and over the next few months they secretly collected two DNA samples from him. One came from his car door, since touching an object often leaves skin cells behind. The other came from a discarded tissue taken from his curbside trash. The DNA was reportedly a perfect match with the killer's. All in all, it was brilliant detective work.

Still, it did raise concerns about genetic privacy. The police probably took DeAngelo's DNA without a warrant. Moreover, DeAngelo's relatives never gave law enforcement permission to use their genetic data. Now, it's hard to have sympathy for alleged serial killers, but the consequences here go far beyond one case. Imagine your mother or sibling or long-lost cousin—someone you've never met—posting their DNA online. Genetic sleuths could now snoop on you and your family, exposing adoptions and affairs in your past and spying on your susceptibility to diseases. Harassment, blackmail, and discrimination are real possibilities. As genetic testing becomes more common, expect to see laws governing who can access such data. Someday, using DNA to expose secrets could be grounds for jail time.

Even for detectives, the ubiquity of genetic technology could end up causing as many problems as it solves. Per the Golden State Killer case, your garbage is full of DNA, mostly from skin cells. In theory, a rogue scientist could harvest, culture, and deprogram those skin cells, turning them back into stem cells. Stem cells can then be converted into any other type of cell in the body, including blood and sperm cells. With a little biological black magic, you'd suddenly have the ability to plant anyone's bodily fluids at any crime scene, either to frame that person or to sow so much doubt that the real killer goes free.

Genetic engineering could also enable dastardly new forms of murder. Aside from identical twins, we all have unique DNA, including unique flaws and vulnerabilities. A clever scientist could therefore design a silver-bullet virus that, even if released in a public place, would target and kill just one person.

We could also bring extinct life forms back from the dead, a morally fraught idea. Consider woolly mammoths. Mammoth bones and pelts abound in Siberia, and the cold climate there preserves mammoth DNA quite well. Imagine splicing that mammoth DNA into the embryo of an elephant, and implanting that embryo into an elephant womb. The resulting calf wouldn't be a pure woolly mammoth. But it would be close, with shaggy fur and curly tusks and several key physiological traits. In a functional sense, then, we could easily resurrect the mammoth from extinction.

But should we? Pachyderms are pack animals—highly smart, highly social. They need companions or they suffer. Eventually, of course, we could raise a whole herd of mammoths for company. But that first mammoth would be intensely lonely, a hell of an awful life. That's assuming the DNA splicing and editing goes smoothly, too, which it probably won't. What if severe birth defects arise? How far are we willing to push things for an experiment?

The moral recoil would be even stronger with Neanderthals. Although they have a reputation in popular culture as brutes, the best archaeological evidence suggests that Neanderthals were every bit as smart as humans. Based on their skull size, they had bigger brains than us. They also made art, played music, crafted tools, buried their dead, and possibly had language. Humans and Neanderthals even interbred in the recent past, so we're quite similar genetically. Just like with mammoths and elephants, then, scientists could splice Neanderthal DNA into a human embryo and implant it in a human womb. Nine months later you'd have, in effect, the first Neanderthal to walk the Earth in 40,000 years.

But even more so than mammoths, Neanderthals were probably highly social, as much so as human beings. We could try raising a Neanderthal child in human society, but would she ever truly fit in? Perhaps she'd always be an "other." Calling this resurrection a *crime* doesn't quite seem right; and philosophically, perhaps it's better to exist than not exist. But it's ethically dubious at best, and if done wrong could be needlessly cruel.

This litany of future crimes isn't meant to be dystopian: None of these potential misdeeds is inevitable. It's important to recognize that we'll of course *benefit* from future technologies, too, often mightily—we'll eliminate diseases, free ourselves from drudgery, open our minds to new horizons, and so on. Moreover, science and technology can also solve and prevent crimes. DNA technology cracks open cold cases. Satellites help archaeologists monitor remote dig sites to cut down on looting, and help aid groups expose human trafficking and modern slavery.*

Some of the crimes mentioned above seem farfetched, to be honest. (Homicidal sexbots?) But the future probably always seems

*An astounding 40 million people around the world are currently enslaved, mostly in the fishing, mining, and brickmaking industries in developing countries. While slave camps can easily elude detection on the ground, they can't hide from satellites. AI algorithms can then learn the distinguishing features of slave camps and swiftly sort through satellite images to pinpoint their locations.

outlandish from a distance. If you'd told someone in 1900 that people today would be using boxes of electrons to steal cash from banks or graft their ex-girlfriends' faces into revenge porn, that would have seemed pretty crazy. Yet here we are. Perhaps the worst crimes will be ones we can't even envision. Imagine all the havoc you could wreak with time travel, or cyborg brains that tapped into supercomputers.

Overall, I hope you found this sketch of future crimes to be thought-provoking—and useful. It's always valuable to think through how people might abuse technologies: we can't safeguard against all evils, but those who unleash new powers into the world have a moral duty, I'd argue, to mitigate what risks they can. I'm sure there are potential dastardly deeds I've overlooked as well. If you can think of any more, please get in touch at samkean.com/contact. Above all, thanks for reading . . .

ACKNOWLEDGMENTS

However fascinating the stories in this book are, this was not always a fun book to write: There was a lot of misery to chronicle, too. I'd therefore like to take a moment to remember all the men and women who suffered for the sake of—and at the hands of—science over the centuries. Science has given us a lot, no question, and scientists should be proud of their record overall. But science can and should do better, and the stories of its victims deserve to be more widely known.

Many more people contribute to the writing of a book than the author, and I couldn't have completed this one without the help of a whole host. There was my steadfast agent Rick Broadhead, who was always there with advice. There was my editor Phil Marino, whose deft suggestions shaped the manuscripts and make it sparkle. There were dozens of other people in and around Little, Brown as well, including Liz Gassman, Deri Reed, and Michael Noon. This book wouldn't be in your hands now without them.

I also owe a big thanks to my friends and family: My parents Jean and Gene, who continue to be my biggest fans and best sales reps. My brother Ben and his partner Nicole in Washington, D.C., who kept me sane during the pandemic with rooftop beers. My sister

Becca and her husband John back in South Dakota, whose boat pictures make me jealous, but whose pictures of Penny and Harry always lift my spirits. (Go Flyers!) And to my new and longtime friends in Washington, D.C., and all around the world—I can't wait to see you all again soon.

As I've said before, a few lines on a page aren't sufficient to express all my gratitude, and if I've left anyone off this list, I remain thankful, if embarrassed . . .

WORKS CITED

Prologue: Cleopatra's Legacy

Cleopatra: A Life, by Stacy Schiff, Back Bay Books, 2011

"Cleopatra's Children's Chromosomes: A *Halachic* Biological Debate," by Merav Gold, accessed on November 15th, 2020, at http://download.yutorah.org/2016/1053/857234.pdf

"The Life of Antony," in *Parallel Lives*, by Plutarch, accessed on November 15th, 2020, at http://penelope.uchicago.edu/Thayer/E/Roman/Texts/Plutarch/Lives/Antony*.html

"Nazi Medical Experimentation: The Ethics Of Using Medical Data From Nazi Experiments," in *The Journal of Halacha and Contemporary Society*, by Baruch Cohen, Spring 1990, issue 19, pages 103-26

Rise of Fetal and Neonatal Physiology: Basic Science to Clinical Care, by Lawrence D. Longo, Springer-Verlag New York, 2013

When Doctors Kill: Who, Why, and How, by Joshua A. Perper and Stephen J. Cina, Copernicus, 2010

Introduction

"Fourteen Psychological Forces That Make Good People Do Bad Things," by Travis Bradberry, last accessed November 19th, 2020, at http://huffpost.com/entry/14-psychological-forces-t_b_9752132

"The Science of Why Good People Do Bad Things," from Psychology Today.com, by Ronald E. Riggio, last accessed November 19th, 2020, at http://psychologytoday.com/us/blog/cutting-edge-leadership/201411/the-science-why-good-people-do-bad-things

"Why Do Good People Do Bad Things?", from Ethics Alliance, by Daniel Effron, August 14th, 2018, last accessed November 19th, 2020, at https://ethics.org.au/good-people-bad-deeds/

"Why Ethical People Make Unethical Choices," in *Harvard Business Review*, by Ron Carucci, December 16th, 2016, last accessed November 19th, 2020, at https://hbr.org/2016/12/why-ethical-people-make-unethical-choices

Chapter 1: Piracy: The Buccaneer Biologist

"Bioprospecting/Biopiracy and Indigenous Peoples," by the ETC Group, December 26th, 1995, accessed at https://www.etcgroup.org/content/bioprospectingbiopiracy-and-indigenous-peoples

"Discourse on Winds," in *Voyages and Descriptions*, by William Dampier, 1699, accessed through Google Books

The Drunken Botanist, by Amy Stewart, Algonquin Books, 2013

The Faces of Crime and Genius: The Historical Impact of the Genius-Criminal, by Dean Lipton, A.S. Barnes & Company, 1970

The Fever Trail: In Search of the Cure for Malaria, by Mark Honigsbaum, Picador, 2003

Global Biopiracy: Patents, Plants, and Indigenous Knowledge, by Ikechi Mgbeoji, Cornell University Press, 2006

Henry Smeathman, the Flycatcher: Natural History, Slavery, and Empire in the Late Eighteenth Century, by Deirdre Coleman, Liverpool University Press, 2018

"Natural History, Improvement, and Colonisation: Henry Smeathman and Sierra Leone in the Late Eighteenth Century," by Starr Douglas, Ph.D. thesis, University of London, available at https://ethos.bl.uk/OrderDetails.do?uin=uk.bl.ethos.409707

New Voyage Around the World, by William Dampier, 1697, available through Google Books

"Perils of Plant Collecting," by A.M. Martin, accessed on November 15th, 2020, at https://web.archive.org/web/20120127142335/https://www.lmi.org.uk/Data/10/Docs/16/16Martin.pdf

Pirate of Exquisite Mind: The Life of William Dampier, by Diana Preston and Michael Preston, Transworld, 2005
Plant Hunters: The Adventures of the World's Greatest Botanical Explorers, by Carolyn Fry, University of Chicago Press, 2013
"A Slaving Surgeon's Collection: The Pursuit of Natural History through the British Slave Trade to Spanish America," in *Curious Encounters Voyaging, Collecting, and Making Knowledge in the Long Eighteenth Century*, by Kathleen S. Murphy, University of Toronto Press, 2019

Chapter 2: Slavery: The Corruption of the Flycatcher

"Collecting Slave Traders: James Petiver, Natural History, and the British Slave Trade," in *William and Mary Quarterly*, by Kathleen S. Murphy, volume 70, issue 4, pages 637–670, October 2013
"Enlightenment, Scientific Exploration and Abolitionism: Anders Sparrman's and Carl Bernhard Wadström's Colonial Encounters in Senegal, 1787–1788 and the British Abolitionist Movement," in *Slavery & Abolition*, by Klas Rönnbäck, volume 34, issue 3, pages 425–445, 2013
Henry Smeathman, the Flycatcher: Natural History, Slavery, and Empire in the Late Eighteenth Century, by Deirdre Coleman, Liverpool University Press, 2018
Interviews with Kathleen Murphy, March and April 2019, conducted by Sam Kean
"The making of scientific knowledge in an age of slavery: Henry Smeathman, Sierra Leone and natural history," in *Journal of Colonialism & Colonial History*, by Starr Douglas, volume 9, issue 3, Winter 2008
"Natural History, Improvement, and Colonisation: Henry Smeathman and Sierra Leone in the Late Eighteenth Century," by Starr Douglas, Ph.D. thesis, University of London, available at https://ethos.bl.uk/OrderDetails.do?uin=uk.bl.ethos.409707
Plan of a Settlement to Be Made Near Sierra Leona on the Grain Coast of Africa, by Henry Smeathman, 1786, last accessed November 18th, 2020, https://digitalcollections.nypl.org/items/c16ace30-ff74-0133-adc4-00505686a51c
"The Royal Society, Slavery, and the Island of Jamaica: 1660–1700," in *The Notes and Records of the Royal Society Journal of the History of Science*, by Mark Govier, volume 53, issue 2, May 22nd, 1999

"Science's debt to the slave trade," in *Science*, by Sam Kean, April 5th, 2019, volume 364, issue 6435, pages 16–20

"Slavery and the Natural World," by the Natural History Museum, in London, last accessed November 18th, 2020, https://www.nhm.ac.uk/discover/slavery-and-the-natural-world.html

"Slavery in the Cabinet of Curiosities: Hans Sloane's Atlantic World," by James Delburgo, British Museum, 2007, last accessed November 19th, 2020, www.britishmuseum.org/PDF/Delbourgo%20essay.pdf

"A Slaving Surgeon's Collection: The Pursuit of Natural History through the British Slave Trade to Spanish America," in *Curious Encounters Voyaging, Collecting, and Making Knowledge in the Long Eighteenth Century*, by Kathleen S. Murphy, University of Toronto Press, 2019

"Some Account of the Termites Which Are Found in Africa and Other Hot Climates," in *Philosophical Transactions of the Royal Society*, by Henry Smeathman, volume 71, 1781, last accessed November 19th, 2020, https://royalsocietypublishing.org/doi/10.1098/rstl.1781.0033

"The South Sea Company and Contraband Trade," in *The American Historical Review*, by Vera Lee Brown, volume 31, issue 4, July 1926, pages 662–678

Chapter 3 — Grave-Robbing: Jekyll & Hyde, Hunter & Knox

"Acromegalic Gigantism, Physicians, and Body Snatching. Past or Present?" in *Pituitary*, by Wouter W. de Herder, volume 15, pages 312–318, 2012

The Anatomy Murders: Being the True and Spectacular History of Edinburgh's Notorious Burke and Hare and of the Man of Science Who Abetted Them in the Commission of Their Most Heinous Crimes, by Lisa Rosner, University of Pennsylvania Press, 2011

Brain, Vision, Memory: Tales in the History of Neuroscience, by Charles Gross, MIT Press, 1998

The Diary of a Resurrectionist, by James Blake Bailey, 1896, available on Google Books

"The Emperor's New Clothes," *Journal of the Royal Society of Medicine*, by Don C. Shelton, volume 103, pages 46–50, 2010

Explorers of the Body, by Steven Lehrer, Doubleday, 1979

Galileo Goes to Jail and Other Myths about Science and Religion, by Ronald L. Numbers (editor), Harvard University Press, 2010

The Knife Man: Blood, Body Snatching, and the Birth of Modern Surgery, by Wendy Moore, Crown, 2006

Leicester Square: Its Associations and Its Worthies, by Tom Taylor, 1874, available through Google Books

The Life of Sir Astley Cooper, by Bransby Blake Cooper, 1843, available on Google Books

A Sense of the World: How a Blind Man Became History's Greatest Traveler, by Jason Roberts, Harper Perennial, 2007

Sites Of Autopsy In Contemporary Culture, by Elizabeth Klaver, SUNY Press, 2005

"William Smellie and William Hunter: Two Great Obstetricians and Anatomists," in *Journal of the Royal Society of Medicine*, by A.D.G. Roberts, T.F. Baskett, A.A. Calder, and S. Arulkumaran, volume 103, pages 205–206, 2010

Chapter 4 — Murder: The Professor and the Janitor

"Anatomy's Use of Unclaimed Bodies: Reasons Against Continued Dependence on an Ethically Dubious Practice," in *Clinical Anatomy*, by D. Gareth Jones and Maja I. Whitaker, volume 25, issue 2, pages 246–254, March 2012

"The Art of Medicine: American Resurrection and the 1788 New York Doctors' Riot," in *The Lancet*, by Caroline de Costa and Francesca Miller, volume 377, issue 9762, pages 292–293, January 22, 2011

"Bill Would Require Relatives' Consent for Schools to Use Cadavers," in *The New York Times*, by Nina Bernstein, June 26th, 2016, last accessed November 21st, 2020, at www.nytimes.com/2016/06/27/nyregion/new-yorks-written-consent-bill-would-tighten-use-of-bodies-for-teaching.html

Blood & Ivy: The 1849 Murder That Scandalized Harvard, by Paul Collins, W.W. Norton, 2018

"A Brief But Sordid History of the Use of Human Cadavers in Medical Education," in *Proceedings of the 13th Annual History of Medicine Days* (W.A. Whitelaw, ed.), by Melanie Shell, Faculty of Medicine, The University of Calgary, 2004

"A Brief History of American Anatomy Riots," from The National Museum of Civil War Medicine, by Bess Lovejoy, last accessed November 21st, 2020, at https://www.civilwarmed.org/anatomy-riots/

"The Doctors Riot 1788," from The History Box, last accessed November 21st, 2020, at http://thehistorybox.com/ny_city/riots/riots_article7a.htm

"The Gory New York City Riot that Shaped American Medicine," from SmithsonianMag.com, by Bess Lovejoy, last accessed November 21st, 2020, at https://www.smithsonianmag.com/history/gory-new-york-city-riot-shaped-american-medicine-180951766/

History of Medicine in New York: Three Centuries of Progress, by James J. Walsh, National Americana Society, 1919

"Human Corpses Are Prize In Global Drive For Profits," from the International Consortium of Investigative Journalists, by By Kate Willson, Vlad Lavrov, Martina Keller, Thomas Maier, and Gerard Ryle, last accessed on November 21st, 2020, at https://www.huffpost.com/entry/human-corpses-profits_b_1679094

"The Janitor's Story: An Ethical Dilemma in the Harvard Murder Case," in the *American Bar Association Journal*, by Albert I. Borowitz, volume 66, issue 12, pages 1540-1545, December 1980

"Murder at Harvard," in *The American Scholar*, by Stewart Holbrook, volume 14, issue 4, pages 425-434, Autumn 1945

Trouble With Testosterone: And Other Essays On The Biology Of The Human Predicament, by Robert Sapolsky, Scribner, 1998

Chapter 5 — Animal Cruelty: War of the Currents

"Five Little Piggies: An Anecdotal Account of the History of the Anti-Vivisection Movement," in *Proceedings of the 10th Annual History of Medicine Days* (W.A. Whitelaw, ed.), by Vicky Houtzager, Faculty of Medicine, The University of Calgary, 2001

"Are animal models predictive for humans?", in *Philosophy, Ethics, and Humanities in Medicine*, by Niall Shanks, Ray Greek, and Jean Greek, volume 4, issue 2, 2009

Auburn Correctional Facility (Images of America), by Eileen McHugh and Cayuga Museum, Arcadia Publishing, 2010

Brain, Vision, Memory: Tales in the History of Neuroscience, by Charles Gross, MIT Press, 1998

"The Dangers of Electric Lighting," *The North American Review*, by Thomas Edison, volume 149, issue 396, pages 625-634, November 1889

Edison and the Electric Chair, by Mark Essig, Walker Books, 2004

"Edison and 'The Chair,'" in *IEEE Technology and Society Magazine*, by Terry S. Reynolds and Theodore Bernstein, volume 8, issue 1, March 1989

The Electric Chair: An Unnatural American History, by Craig Brandon, McFarland, 2009

"Electrifying Story," in *The Threepenny Review*, by Arthur Lubow, issue 49, pages 31-32, spring 1992

Empires of Light: Edison, Tesla, Westinghouse, and the Race to Electrify the World, by Jill Jonnes, Random House, 2004

"Harold P. Brown and the Executioner's Current: An Incident in the AC-DC Controversy," in *The Business History Review*, by Thomas P. Hughes, volume 32, issue 2, pages 143–165, summer 1958

Henry Smeathman, the Flycatcher: Natural History, Slavery, and Empire in the Late Eighteenth Century, by Deirdre Coleman, Liverpool University Press, 2018

"Heroes, Herds, and Hysteresis in Technological History: Thomas Edison and 'The Battle of the Systems' Reconsidered," *Industrial and Corporate Change*, by Paul A. David, volume 1, issue 1, pages 129–180, 1992

"'Killing the Elephant': Murderous Beasts and the Thrill of Retribution, 1885–1930," in *The Journal of the Gilded Age and Progressive Era*, by Amy Louise Wood, volume 11, issue 3, pages 405–444, July 2012

The Knife Man: Blood, Body Snatching, and the Birth of Modern Surgery, by Wendy Moore, Crown, 2006

"Life and Death by Electricity in 1890: The Transfiguration of William Kemmler," in *Journal of American Culture*, by Nicholas Ruddick, volume 21, issue 4, pages 79–87, Winter 1998

"Modern biomedical research: an internally self-consistent universe with little contact with medical reality?", in *Nature Reviews*, by David F. Horrobin, volume 2, February 2003, pages 151–154

"Natural History, Improvement, and Colonisation: Henry Smeathman and Sierra Leone in the Late Eighteenth Century," by Starr Douglas, Ph.D. thesis, University of London, available at https://ethos.bl.uk/OrderDetails.do?uin=uk.bl.ethos.409707

Neurotribes: The Legacy of Autism and the Future of Neurodiversity, by Steve Silberman, Avery 2016

The Power Makers, by Maury Klein, Bloomsbury, 2008

Racial Hygiene: Medicine under the Nazis, by Robert N. Proctor, Harvard University Press, 1990

"Mr. Brown's Rejoinder," in *The Electrical Engineer*, volume 7, pages 369–370, August 1888

Topsy: The Startling Story of the Crooked Tailed Elephant, P. T. Barnum, and the American Wizard, Thomas Edison, by Michael Daly, Atlantic Monthly Press, 2013

"Is the Use of Sentient Animals in Basic Research Justifiable?" in *Philosophy, Ethics, and Humanities in Medicine*, by Ray Greek and Jean Greek, volume 5, issue 14, 2010

Chapter 6 — Sabotage: The Bone Wars

Beasts of Eden: Walking Whales, Dawn Horses, and Other Enigmas of Mammal Evolution, by David Rains Wallace, University of California Press, 2004

"Bone Wars: The Cope-Marsh Rivalry," from The Academy of Natural Sciences, last accessed on November 21st, 2020, at https://ansp.org/exhibits/online-exhibits/stories/bone-wars-the-cope-marsh-rivalry/

The Bonehunters' Revenge: Dinosaurs and Fate in the Gilded Age, by David Rains Wallace, Mariner Books, 2000

Dinosaurs in the Attic: An Excursion into the American Museum of Natural History, by Douglas J. Preston, St. Martin's Press, 2014

"Edward Drinker Cope's final feud," in *Archives of Natural History*, by P. D. Brinkman, volume 43, issue 2, pages 305–320, 2016

"Empire and Extinction: The Dinosaur as a Metaphor for Dominance in Prehistoric Nature," in *Leonardo*, by Paul Semonin, volume 30, issue 3, pages 171–182, 1997

The Gilded Dinosaur: The Fossil War Between E.D. Cope and O.C. Marsh and the Rise of American Science, by Mark Jaffe, Crown, 2000

The Great Dinosaur Hunters and Their Discoveries, by Edwin H. Colbert, Dover, 1984

"Marsh Hurles Azoic Facts at Cope," in *New York Herald*, by William Hosea Ballou, January 19th, 1890, page 11

"Professor Cope Vs. Professor March," in *American Heritage*, by James Penick Jr., volume 22, issue 5, August 1971

"Remarking on a Blackened Eye: Persifor Frazer's Blow-by-Blow Account of a Fistfight with His Dear Friend Edward Drinker Cope," in *Endeavour*, by Paul D. Brinkman, volume 39, issue 3–4, pages 188–192, Sept.-Dec. 2015

"Scientists Wage Bitter Warfare," in *New York Herald*, by William Hosea Ballou, January 21st, 1890, page 10–11

Some Memories of a Paleontologist, by William Berryman Scott, Princeton University Press, 1939

"The Uintatheres and the Cope-Marsh War," in *Science*, by Walter H. Wheeler, volume 131, issue 3408, pages 1171–1176, April 22nd, 1960

"Volley for Volley in the Great Scientific War," in *New York Herald*, by William Hosea Ballou, January 13th, 1890, page 4

Chapter 7 — Oath-Breaking: Ethically Impossible

"Anti-Smoking Initiatives in Nazi Germany: Research and Public Policy," in *Proceedings of the 11th Annual History of Medicine Days* (W.A. Whitelaw, ed.), by Nathaniel Dostrovsky, Faculty of Medicine, The University of Calgary, 2002

Asperger's Children: The Origins of Autism in Nazi Vienna, by Edith Sheffer, W. W. Norton, 2018

"Can Evil Beget Good? Nazi Data: A Dilemma for Science," in the *Los Angeles Times*, Barry Siegel, October 30th, 1998, page 1

"Eponyms and the Nazi Era: Time to Remember and Time for Change," in the *Israel Medical Association Journal*, by Rael D. Strous and Morris C. Edelman, volume 9, issue 3, pages 207–214, March 2007

"Ethical Complexities of Conducting Research in Developing Countries," in the *New England Journal of Medicine*, by Harold Varmus, M.D., and David Satcher, volume 337, pages 1003-1005

"Ethical Dilemmas with the Use of Nazi Medical Research," in *Proceedings of the 11th Annual History of Medicine Days* (W.A. Whitelaw, ed.), by Batya Grundland and Eve Pinchefsky, Faculty of Medicine, The University of Calgary, 2001

"Ethical Failures and History Lessons: The U.S. Public Health Service Research Studies in Tuskegee and Guatemala," in *Public Health Reviews*, by Susan M. Reverby, volume 34, issue 13, 2012

"The Ethical Use of Unethical Human Research," by Jonathan Steinberg, last accessed on November 21st, 2020, at http://www.bioethics.as.nyu.edu/docs/IO/30171/Steinberg.HumanResearch.pdf

"'Ethically Impossible': STD Research in Guatemala from 1946 to 1948," from The Presidential Commission for the Study of Bioethical Issues, September 2011, last accessed on November 21st, 2020, at https://bioethicsarchive.georgetown.edu/pcsbi/node/654.html

"Ethically Sound: Ethically Impossible," the Ethically Sound podcast, from the Presidential Commission for the Study of Bioethical Issues, last accessed on November 21st, 2020, at https://bioethicsarchive.georgetown.edu/pcsbi/node/5896.html

Examining Tuskegee: The Infamous Syphilis Study and Its Legacy, by Susan M. Reverby, University of North Carolina Press, 2013

"Exposed: US Doctors Secretly Infected Hundreds of Guatemalans with Syphilis in the 1940s," from Democracy Now, last accessed on November 21st, 2020, at https://www.democracynow.org/2010/10/5/exposed_us_doctors_secretly_infected_hundreds

"The Guatemala Experiments," in *Pacific Standard Magazine*, by Mike Mariani, last accessed November 21st, 2020, at https://psmag.com/news/the-guatemala-experiments

The Knife Man: Blood, Body Snatching, and the Birth of Modern Surgery, by Wendy Moore, Crown, 2006

"Linking Groupthink to Unethical Behavior in Organizations," in *Journal of Business Ethics*, by Ronald R. Sims, volume 11, pages 651–662, 1992

"Nazi Medical Experimentation: The Ethics Of Using Medical Data From Nazi Experiments," in *The Journal of Halacha and Contemporary Society*, by Baruch Cohen, Spring 1990, issue 19, pp. 103–26

"Nazi Hypothermia Research: Should the Data Be Used?", *Military Medical Ethics*, Volume 2, by Robert S. Pozos, last accessed on November 21st, 2020, at https://ke.army.mil/bordeninstitute/published_volumes/ethicsVol2/Ethics-ch-15.pdf

Neurotribes: The Legacy of Autism and the Future of Neurodiversity, by Steve Silberman, Avery 2016

"'Normal Exposure' and Inoculation Syphilis: A PHS "Tuskegee" Doctor in Guatemala, 1946–1948," in *Journal of Policy History*, by Susan Reverby, volume 23, issue 1, 2011, pages 6–28

"Obituary: John Charles Cutler / Pioneer in preventing sexual diseases," in *The Pittsburgh Post-Gazette*, by Jan Ackerman, February 12th, 2003, last accessed on November 21st, 2020, at https://old.post-gazette.com/obituaries/20030212cutler0212p3.asp

"On the Philosophical and Historical Implications of the Infamous Tuskegee Syphilis Trials," in *Proceedings of the 11th Annual History of Medicine Days* (W.A. Whitelaw, ed.), by Tomas Jiminez, Faculty of Medicine, The University of Calgary, 2002

Operation Paperclip: The Secret Intelligence Program that Brought Nazi Scientists to America, by Annie Jacobsen, Back Bay Books, 2015

Racial Hygiene: Medicine under the Nazis, by Robert N. Proctor, Harvard University Press, 1990

"Reflections on the Inoculation Syphilis Studies in Guatemala," Agents of Change podcast, from Lehman University, transcript last accessed on November 21st, 2020, at http://wp.lehman.edu/lehman-today/reflections-on-the-inoculation-syphilis-studies-in-guatemala/

"Results of Death-Camp Experiments: Should They Be Used? All 14 Counterarguments," from PBS NOVA, last accessed on November 21st, 2020, at https://www.pbs.org/wgbh/nova/holocaust/experifull.html

The Science of Evil: On Empathy and the Origins of Cruelty, by Simon Baron-Cohen, Basic Books, 2012

"Thirty Neurological Eponyms Associated with the Nazi Era," in *European Neurology*, by Daniel Kondziella, volume 62, issue 1, pages 56–64, 2009

"The Treatment of Shock from Prolonged Exposure to Cold, Especially in Water," from Allied Forces, Supreme Headquarters, Combined Intelligence Objectives, by Leo Alexander, last accessed on November 21st, 2020, at https://collections.nlm.nih.gov/catalog/nlm:nlmuid-101708929-bk

"The Victims of Unethical Human Experiments and Coerced Research under National Socialism," in *Endeavour*, by Paul Weindling, Anna von Villiez, Aleksandra Loewenau, Nichola Farron, volume 40, issue 1, 2015

"Why Did So Many German Doctors Join the Nazi Party Early?", in *International Journal of Law and Psychiatry*, by Omar S. Haque, Julian De Freitas, Ivana Viani, Bradley Niederschulte, Harold J. Bursztajn, volume 35, issues 5–6, pages 473–479, 2012

"WHO's malaria vaccine study represents a 'serious breach of international ethical standards,'" in *The British Medical Journal*, by Peter Doshi, volume 268, pages 734–735

Chapter 8 — Ambition: Surgery of the Soul

"Fighting the Legend of the 'Lobotomobile,'" by Jack El-Hai, from Wonders & Marvels, last accessed on November 21st, 2020, at https://www.wondersandmarvels.com/2016/03/fighting-the-legend-of-the-lobotomobile.html

Great and Desperate Cures: The Rise and Decline of Psychosurgery and Other Radical Treatments for Mental Illness, by Elliot S. Valenstein, Basic Books, 1986

The Great Pretender: The Undercover Mission That Changed Our Understanding of Madness, by Susannah Cahalan, Grand Central Publishing, 2019

The Lobotomist: A Maverick Medical Genius and His Tragic Quest to Rid the World of Mental Illness, by Jack El-Hai, Wiley, 2007

An Odd Kind of Fame: Stories of Phineas Gage, by Malcolm Macmillan, The MIT Press, 2000

"The Operation of Last Resort," *The Saturday Evening Post*, by Irving Wallace, October 20, 1951, pages 24–25, 80, 83–84, 89–90, 92, 94–95

Ten Drugs: How Plants, Powders, and Pills Have Shaped the History of Medicine, by Thomas Hager, Harry N. Abrams, 2019

Chapter 9 — Espionage: The Variety Act

Bombshell: The Secret Story of America's Unknown Atomic Spy Conspiracy, by Joseph Albright and Marcia Kunstel, Times Books, 1997

The Brother: The Untold Story of the Rosenberg Case, by Sam Roberts, Simon & Schuster, 2014

Cannibalism: A perfectly natural history, by Bill Schutt, Algonquin, 2017

Dark Sun: The Making of the Hydrogen Bomb, by Richard Rhodes, Simon & Schuster, 1996

"Extracts From Testimony Given by Harry Gold at Spy Trial," in *The New York Times*, March 16, 1951, page 9

The FBI-KGB War: A Special Agent's Story, by Robert J. Lamphere, Random House, 1986

Food and Famine in the 21st Century, by William A. Dando, ABC-CLIO, 2012

"Harry Gold: Spy in the Lab," in *Distillations*, by Sam Kean, last accessed November 22, 2020, at https://www.sciencehistory.org/distillations/harry-gold-spy-in-the-lab

Hungry Ghosts: Mao's Secret Famine, by Jasper Becker, 2013

Invisible Harry Gold: The Man Who Gave the Soviets the Atom Bomb, by Allen M. Hornblum, Yale University Press, 2010

Klaus Fuchs, Atom Spy, by Robert Chadwell Williams, Harvard University Press, 1987

"Lysenko Rising," in *Current Biology*, by Florian Maderspacher, volume 20, issue 19, pages R835–R836, October 12th, 2010

Lysenko's Ghost: Epigenetics and Russia, by Loren Graham, Harvard University Press, 2016

Racial Hygiene: Medicine under the Nazis, by Robert N. Proctor, Harvard University Press, 1990

Red Spies in America: Stolen Secrets and the Dawn of the Cold War, by Katherine A.S. Sibley, University Press of Kansas, 2004

"Rethinking Lysenko's Legacy," in *Science*, by Maurizio Meloni, volume 352, issue 6284, page 421

"Russia's New Lysenkoism," in *Current Biology*, by Edouard I. Kolchinsky, Ulrich Kutschera, Uwe Hossfeld, and Georgy S. Levit, volume 27, issue 19, pages R1042–R1047, October 9th, 2017

"Soviet Atomic Espionage," Joint Committee on Atomic Energy, hearings on Soviet Atomic Energy, April 1951, Printed for the use of the Joint Committee on Atomic Energy, Government Printing Office, last accessed on November 21st, 2020, at https://archive.org/stream/sovietatomicespi1951unit/sovietatomicespi1951unit_djvu.txt

"The Soviet Union's Scientific Marvels Came from Prisons," from *The Atlantic*, by Marina Koren, published May 5th, 2017, last accessed on November 28th, 2020, at https://www.theatlantic.com/science/archive/2017/05/soviet-science-stalin/525576/

The Spy Who Changed The World, by Mike Rossiter, Headline, 2015

Stalin and the Bomb: Soviet Union and Atomic Energy, 1939-56, by David Holloway, Yale University Press, 1994

"Stalin's War on Genetics," in *Nature*, by Jan Witkowski, volume 454, issue 7204, pages 577–579, July 31st, 2008

"Testimony of Harry Gold," from the Department of Justice, Office of the U.S. Attorney for the Southern Judicial District of New York, last accessed on November 22nd, 2020, at https://catalog.archives.gov/id/2538330

Venona: Decoding Soviet Espionage in America, by John Earl Haynes and Harvey Klehr, Yale University Press, 2000

*The Venona S*ecrets: The Definitive Exposé of Soviet Espionage in America, by Herbert Romerstein and Eric Breindel, Regnery History, 2014

Chapter 10 — Torture: The White Whale

The Big Test: The Secret History of the American Meritocracy, by Nicholas Lemann, 2000, Farrar, Straus, and Giroux

Blood & Ivy: The 1849 Murder That Scandalized Harvard, by Paul Collins, W.W. Norton, 2018

"Buying a Piece of Anthropology: Part One: Human Ecology and unwitting anthropological research for the CIA," in *Anthropology Today*, by David H. Price, volume 23, issue 3, pages 8–13, June 2007

"Buying a Piece of Anthropology: Part Two: The CIA and Our Tortured Past," in *Anthropology Today*, by David H. Price, volume 23, issue 5, pages 17–22, October 2007

"Comparing Soviet and Chinese Political Psychiatry," in *The Journal of the American Academy of Psychiatry and the Law*, by Robert van Voren, volume 30, issue 1, pages 131–135, 2002

Criminal Genius: A Portrait of High-IQ Offenders, by James C. Oleson, University of California Press, 2016

Every Last Tie: The Story of the Unabomber and His Family, by David Kaczynski, Duke University Press, 2016

"Forensic Linguistics, the Unabomber, and the Etymological Fallacy," from Language Log, by Benjamin Zimmer, January 14th, 2006, last accessed on November 22nd, 2020, at itre.cis.upenn.edu/~myl/languagelog/archives/002762.html

Harvard and the Unabomber: The Education of an American Terrorist, by Alston Chase, W.W. Norton, 2003

"Henry A. Murray: Brief life of a personality psychologist: 1893-1988," in *Harvard Magazine*, by Marshall J. Getz, March-April 2014

"Henry A. Murray: The Making of a Psychologist?" in *American Psychologist*, by Rodney G. Triplet, volume 47, issue 2, pages 299–307, February 1992

"Henry A. Murray's Early Career: A Psychobiographical Exploration," in *Journal of Personality*, by James William Anderson, volume 56, issue 1, March 1998

Hunting the Unabomber: The FBI, Ted Kaczynski, and the Capture of America's Most Notorious Domestic Terrorist, by Lis Wiehl and Lisa Pulitzer, Thomas Nelson, 2020

"Origins of the Psychological Profiling of Political Leaders: The US Office of Strategic Services and Adolf Hitler," in *Intelligence and National Security*, by Stephen Benedict Dyson, volume 29, issue 5, 654–674, 2014

"Political Abuse of Psychiatry—An Historical Overview," in *Schizophrenia Bulletin*, by Robert van Voren, volume 36, issue 1, pages 33–35, 2010

"Political Abuse of Psychiatry in Authoritarian Systems," in *Irish Journal of Psychological Medicine*, by J. P. Tobin, volume 30, pages 97–102, 2013

"Political Abuse of Psychiatry in the Soviet Union and in China: Complexities and Controversies," in *The Journal of the American Academy of Psychiatry and the Law*, by Richard J. Bonnie, volume 30, issue 1, pages 136–144, 2002

"Political Abuse of Psychiatry with a Special Focus on the USSR," in *The Bulletin of the Royal College of Psychiatrists*, by James Finlayson, volume 11, issue 4, pages 144–145, April 1987

"Portrait: Henry A. Murray," in *The American Scholar*, by Hiram Haydn, volume 39, issue 1, pages 123–136, Winter 1969–70

"Prisoner of Rage: From a Child of Promise to the Unabom Suspect," in *The New York Times*, by Robert D. McFadden, May 26, 1996, last accessed November 22nd, 2020, at nytimes.com/1996/05/26/us/prisoner-of-rage-a-special-report-from-a-child-of-promise-to-the-unabom-suspect.html

"Project MK-ULTRA, The CIA's Program Of Research In Behavioral Modification," Joint Hearing Before the Select Committee on Intelligence and the Subcommittee on Health and Scientific Research of the Committee on Human Resources, United States Senate, 95th Congress, First Session, August 3rd, 1977, U.S. Government Printing Office, 1977, 052-070-04357-1

"Reading the Wounds," in *Search*, by Jina Moore, November/December 2008, pages 26–33

The Science of Evil: The Science of Evil: On Empathy and the Origins of Cruelty, by Simon Baron-Cohen, Basic, 2012

The Search for the Manchurian Candidate, The CIA and Mind Control, by John Marks, W. W. Norton, 1991

"A Severed Head, Two Cops, and the Radical Future of Interrogation," from *Wired*, by Robert Kolker, last accessed on November 22nd, 2020, at https://www.wired.com/2016/05/how-to-interrogate-suspects/

"Soviet Psychiatry in the Cold War Era: Uses and Abuses," in *Proceedings of the 10th Annual History of Medicine Days* (W.A. Whitelaw, ed.), by Nathan Kolla, Faculty of Medicine, The University of Calgary, 2001, pages 254-258

"Studies of Stressful Interpersonal Disputations," in *American Psychologist*, by Henry A. Murray, volume 18, issue 1, pages 28–36, 1963

"Toward a Science of Torture?" in *Texas Law Review*, by Gregg Bloche, volume 95, issue 6, pages 1329–1355, 2017

"The Trouble with Harry," in *The American Scholar*, by Paul Roazen, volume 62, issue 2, pages 306, 308, 310-312, Spring 1993

"The World of Soviet Psychology," in *The New York Times Magazine*, by Walter Reich, January 30th, 1983, last accessed on November 22nd, 2020, at www.nytimes.com/1983/01/30/magazine/the-world-of-soviet-psychiatry.html

Chapter 11 — Malpractice: Sex, Power, and Money

"Ablatio penis: Normal Male Infant Sex-Reassigned as a Girl," in *Archives of Sexual Behavior*, by John Money, volume 4, issue 1, 65–71, 1975

"Am I My Brain or My Genitals? A Nature-Culture Controversy in the Hermaphrodite Debate from the mid-1960s to the late 1990s," in *Gesnerus*, by Cynthia Kraus, volume 68, issue 1, pages 80–106, 2011

"Are hormones a 'female problem' for animal research?," in *Science*, by Rebecca M. Shansky, volume 364, issue 6443, pages 823–826, May 31st, 2019

As Nature Made Him: The Boy Who Was Raised As A Girl, by John Colapinto, Harper Perennial, 2006

"The Biopolitical Birth of Gender: Social Control, Hermaphroditism, and the New Sexual Apparatus," in *Alternatives: Global, Local, Political: Biopolitics beyond Foucault*, by Jemima Repo, volume 38, issue 3, pages 228–244, August 2013

"Body Politics," in *The Washington Post*, by Chris Bull, April 30th, 2000 last accessed on November 23rd, 2020, at https://www.washingtonpost.com/archive/entertainment/books/2000/04/30/body-politics/4d3e07d3-0d74-488d-929d-b2b5f2b3d98d/

"The Contributions of John Money: A Personal View," in *The Journal of Sex Research*, by Vern L. Bullough, volume 40, issue 3, pages 230–236, August 2003

"David and Goliath: Nature Needs Nurture," chapter six of *A First Person History of Pediatric Psychoendocrinology*, by John Money, Springer 2002

"David Reimer's Legacy: Limiting Parental Discretion," in *Cardozo Journal of Law & Gender*, by Hazel Glenn Beh and Milton Diamond, volume 12, issue 1, pages 5–30, 2005

"The Five Sexes, Revisited," in *Sciences*, by Anne Fausto-Sterling, volume 40, issue 4, pages 18–23, July-August 2000

"Gender Gap," in *Slate*, by John Colapinto, published June 3rd, 2004, last accessed on November 23rd, 2020, at slate.com/technology/2004/06/why-did-david-reimer-commit-suicide.html

"Intersexuality and the Categories of Sex," in *Hypatia*, by Georgia Warnke, volume 16, issue 3, pages 126–137, Summer 2001

"Intersexuals Struggle to Find Their Identity," in *The Bergen County Record*, by Ruth Padawer, July 25th, 2004, page A1

The Man Who Invented Gender: Engaging the Ideas of John Money, by Terry Goldie, UBC Press, 2014

"Sex Reassignment at Birth: Long-term Review and Clinical Implications," in *Archives of Pediatric Adolescent Medicine*, by Milton Diamond and Keith H. Sigmundson, volume 151, issue 3, pages 298–304, March 1997

"The Sexes: Biological Imperatives," in *Time*, page 34, Monday, January 8th, 1973

"Sexual Identity, Monozygotic Twins Reared in Discordant Sex Roles and a BBC Follow-Up," in *Archives of Sexual Behavior*, by Milton Diamond, volume 11, issue 2, pages 181–185

"'An Unnamed Blank That Craved a Name': A Genealogy of Intersex as Gender," in *Signs [Sex: A Thematic Issue]*, by David A. Rubin, volume 37, issue 4, pages 883–908, Summer 2012

"What Did it Mean To Be a Castrato?", from Gizmodo.com, by Esther Inglis-Arkell, September 24th, 2015, last accessed on November 23rd, 2020, at io9.gizmodo.com/what-did-it-mean-to-be-a-castrato-1732742399

Chapter 12 — Fraud: Superwoman

"21,500 Cases Dismissed due to Forensic Chemist's Misconduct," in *Chemistry World*, by Rebecca Trager, April 25th, 2017, last accessed November 22nd, 2020, at /www.chemistryworld.com/news/21500-cases-dismissed-due-to-forensic-chemists-misconduct/3007173.article

"Annie Dookhan Pursued Renown along a Path of Lies," in *The Boston Globe*, by Sally Jacobs, February 3rd, 2013, last accessed November 22nd, 2020, at https://www.bostonglobe.com/metro/2013/02/03/chasing-renown-path-paved-with-lies/Axw3AxwmD33lRwXatSvMCL/story.html

Betrayers of Truth: Fraud and Deceit in the Halls of Science, by William Broad and Nicholas Wade, Century, 1983

"Chemist Built Up Ties to Prosecutors," *The Boston Globe*, by Andrea Estes and Scott Allen, December 21st, 2012, page A1

"The Chemists and the Cover-Up," in *Reason*, by Shawn Musgrave, March 2019 issue, last accessed November 22nd, 2020, at https://reason.com/2019/02/09/the-chemists-and-the-cover-up/

"Confrontation at the Supreme Court," in *The Texas Journal on Civil Liberties & Civil Rights*, by Olivia B. Luckett, volume 21, issue 2, pages 219–243, Spring 2016

"Confronting Science: Melendez-Diaz and the Confrontation Clause of the Sixth Amendment," in *The FBI Law Enforcement Bulletin*, by Craig C. King, volume 79, issue 8, pages 24–32, August 2010

"Crime labs under the microscope after a string of shoddy, suspect and fraudulent results," in *The America Bar Association Journal*, by Mark Hansen, September 6, 2013, last accessed on November 22nd, 2020, at https://www.abajournal.com/news/article/crime_labs_under_the_microscope_after_a_string_of_shoddy_suspect

Criminal Genius: A Portrait of High-IQ Offenders, by James C. Oleson, University of California Press, 2016

"The Final Tally Is In: Cases in Annie Dookhan Drug Lab Scandal Set for Dismissal, County by County," from MassLive.com, by Gintautas Dumcius, April 19th, 2017, last accessed November 22nd, 2020, at https://www.masslive.com/news/2017/04/the_final_tally_is_in_cases_in.html

"Forensics in Crisis," in *Chemistry World*, by Rebecca Trager, June 15th, 2018, last accessed November 22nd, 2020, at https://www.chemistryworld.com/features/forensics-in-crisis/3009117.article

"Former State Chemist Arrested in Drug Scandal," in *The Boston Globe*, by Milton J. Valencia and John R. Ellement, September 29th, 2012, page A1

"Hard Questions after Litany of Forensic Failures at U.S. Labs," in *Chemistry World*, by Rebecca Trager, December 1st, 2014, last accessed November 22nd, 2020, chemistryworld.com/news/hard-questions-after-litany-of-forensic-failures-at-us-labs/8030.article

"How a Chemist Dodged Lab Protocols," in *The Boston Globe*, by Kay Lazar, September 30th, 2012, page A1

"How Forensic Lab Techniques Work," from HowStuffWorks.com, by Stephanie Watson, last accessed on November 23rd, 2020, at science .howstuffworks.com/forensic-lab-technique2.htm

"I Messed Up Bad: Lesson on the Confrontation Clause from the Annie Dookhan Scandal," in *Arizona Law Review*, by Sean K. Driscoll, volume 56, issue 3, pages 707–740, 2014

"Identification of Individuals Potentially Affected by the Alleged Conduct of Chemist Annie Dookhan at the Hinton Drug Laboratory: Final Report to Governor Deval Patrick," by David E. Meier, Special Counsel to the Governor's Office, August 2013

"Interview Summary of Annie Dookhan," Massachusetts state police reports, last accessed on November 22nd, 2020, at http://www.documentcloud.org/documents/700555-dookhan-interviews-all.html

"Into the Rabbit-Hole: Annie Dookhan Confronts Melendez-Diaz," in *New England Journal on Criminal & Civil Confinement*, by Anthony Del Signore, volume 40, issue 1, 161–190, Winter 2014

"Investigation of the Drug Laboratory at the William A. Hinton State Laboratory Institute, 2002–2012," from the office of Glenn A. Cunha, Inspector General, Office of the Inspector General, Commonwealth of Massachusetts, March 4th, 2014

"Melendez-Diaz, One Year Later," in *The Boston Bar Journal*, by Martin F. Murphy and Marian T. Ryan, volume 54, issue 4, Fall 2010

"The National Academy of Sciences Report on Forensic Sciences: What It Means for the Bench and Bar," in *Jurimetrics*, by Harry T. Edwards, volume 51, issue 1, pages 1-15, Fall 2010

"Scientific Integrity in the Forensic Sciences: Consumerism, Conflicts of Interest, and Transparency," in *Science & Justice*, by Nicholas V. Passalacqua, Marin A. Pilloud, and William R. Belcher, volume 59, issue 5, pages 573–579, September 2019

"Surrogate Testimony After Williams: A New Answer to the Question of Who May Testify Regarding the Contents of a Laboratory Report," in *Indiana Law Journal*, by Jennifer Alberts, volume 90, issue 1, Winter 2015

"Throwing out Junk Science: How a New Rule of Evidence Could Protect a Criminal Defendant's Right to Confront Forensic Scientists," in *Journal of Law and Policy*, by Michael Luongo, volume 27, issue 1, pages 221-256, Fall 2018

"Trial by Fire," in *The New Yorker*, by David Grann, September 7th, 2009, last accessed November 22nd, 2020, at https://www.newyorker.com/magazine/2009/09/07/trial-by-fire

"Two More Problems and Too Little Money: Can Congress Truly Reform Forensic Science?," in *Minnesota Journal of Law, Science, and Technology*, by Eric Maloney, volume 14, issue 2, pages 923–949, 2013

"What a Massive Database of Retracted Papers Reveals about Science Publishing's 'Death Penalty,'" from *Science*, by Jeffrey Brainard and Jia You, published October 25th, 2018, last accessed on November 23rd, 2020, at https://www.sciencemag.org/news/2018/10/what-massive-database-retracted-papers-reveals-about-science-publishing-s-death-penalty

"With More Work, Less Time, Dookhan's Tests Got Faster," from WBUR, by Chris Amico, last accessed November 22nd, 2020, at badchemistry.legacy.wbur.org/2013/05/15/annie-dookhan-drug-testing-productivity

Conclusion

"Fourteen Psychological Forces That Make Good People Do Bad Things," by Travis Bradberry, last accessed November 19th, 2020, at http://huffpost.com/entry/14-psychological-forces-t_b_9752132

"The Science of Why Good People Do Bad Things," from PsychologyToday.com, by Ronald E. Riggio, last accessed November 19th, 2020, at http://psychologytoday.com/us/blog/cutting-edge-leadership/201411/the-science-why-good-people-do-bad-things

"Signing at the Beginning Makes Ethics Salient and Decreases Dishonest Self-Reports in Comparison to Signing at the End," in *The Proceedings of the National Academy of Sciences*, by Lisa L. Shu, Nina Mazar, Francesca Gino, Dan Ariely, and Max H. Bazerman, volume 109, issue 108, pages 15197–15200, September 18, 2012

"Why Do Good People Do Bad Things?", from Ethics Alliance, by Daniel Effron, August 14th, 2018, last accessed November 19th, 2020, at https://ethics.org.au/good-people-bad-deeds/

"Why Ethical People Make Unethical Choices," in Harvard Business Review, by Ron Carucci, December 16th, 2016, last accessed November 19th, 2020, at https://hbr.org/2016/12/why-ethical-people-make-unethical-choices

Appendix

"Adversarial Attacks on Medical AI: A Health Policy Challenge," in *Science*, by Samuel G. Finlayson, John D. Bowers, Joichi Ito, Jonathan L. Zittrain, Andrew L. Beam, Isaac S. Kohane, volume 363, issue 6433, pages 1287–1289, March 22nd, 2019

"Can the U.S. Annex the Moon?" in *Slate*, by Christopher Mellon and Yuliya Panfil, published July 8th, 2019, last accessed on November 24th, 2020, at slate.com/technology/2019/07/un-outer-space-treaty-1967-allowed-property.html

"A Complete Guide to Cooking in Space," from Gizmodo.com, by Ria Misra, published April 24th, 2014, last accessed on November 24th, 2020, at io9.gizmodo.com/what-happens-when-you-cook-french-fries-in-space-1566973977

"Crime: Moon Court," transcript from Flash Forward, by Rose Eveleth, published September 10th, 2019, last accessed on November 28th, 2020, at https://www.flashforwardpod.com/2019/09/10/crime-moon-court/

"Did Astronaut Lisa Nowak, Love Triangle Attacker, Wear A Diaper?" from ABCNews.com, by Eric M. Strauss, published February 16th, 2011, last accessed on November 28th, 2020, at https://abcnews.go.com/TheLaw/astronaut-love-triangle-attacker-lisa-nowak-wear-diaper/story?id=12932069

"Do Some Surgical Implants Do More Harm Than Good?" in *The New Yorker*, by Jerome Groopman, April 20th, 2020, last accessed on November 28th, 2020, at newyorker.com/magazine/2020/04/20/do-some-surgical-implants-do-more-harm-than-good

"Everything You Never Thought to Ask About Astronaut Food," from *The Atlantic*, by Marina Koren, December 15th, 2017, last accessed on November 24th, 2020, at theatlantic.com/science/archive/2017/12/astronaut-food-international-space-station/548255/

"FBI Agents To Visit Antarctica In Rare Investigation Of Assault," in *The Spokane Spokesman-Review*, by Peter James Spielmann, published October 14th, 1996, last accessed on November 27th, 2020, at https://www.spokesman.com/stories/1996/oct/14/fbi-agents-to-visit-antarctica-in-rare/

"A First Look at the Crypto-Mining Malware Ecosystem: A Decade of Unrestricted Wealth," from arXiv.org, by Sergio Pastrana and

Guillermo Suarez-Tangil, published on September 25th, 2019, last accessed on November 24th, 2020, at https://arxiv.org/pdf/1901.00846.pdf

"Former Astronaut Lisa Nowak's Navy Career Is Over," from Space.com, published August 20th, 2010, last accessed on November 28th, 2020, at https://www.space.com/8990-astronaut-lisa-nowak-navy-career.html

"The Great NASA Bake-Off," from *The Atlantic*, by Marina Koren, published August 3rd, 2019, last accessed on November 25th, 2020, at https://www.theatlantic.com/science/archive/2019/08/cookies-in-space/595396/

"History Lessons for Space," in *Slate*, by Russell Shorto, published, July 4th, 2010, last accessed on November 25th, 2020, at https://slate.com/technology/2019/07/manhattan-new-amsterdam-history-settling-space.html

"History of Space Medicine: A North American Perspective," in *Proceedings of the 10th Annual History of Medicine Days* (W.A. Whitelaw, ed.), by Nishi Rawat, Faculty of Medicine, The University of Calgary, 2001

The Horizontal Everest: Extreme Journeys on Ellesmere Island, by Jerry Kobalenko, Soho Press, 2002

"Houston, We Have a Bake-Off! We Finally Know What Happens When You Bake Cookies in Space," from Space.com, by Chelsea Gohd, published January 24th, 2020, last accessed on November 28th, 2020, at https://www.space.com/first-space-cookies-final-baking-results-aroma.html

"How Weird Is It That a Company Lost Hundreds of Millions in Cryptocurrency Because Its CEO Died?" in *Slate*, by Aaron Mak, published December 18th, 2019, last accessed on November 28th, 2020, at https://slate.com/technology/2019/12/quadriga-gerald-cotten-death-cryptocurrency.html

"How Will People Behave in Deep Space Disasters?" in *Slate*, by Amanda Ripley, published May 25th, 2019, last accessed on November 25th, 2020, at slate.com/technology/2019/05/space-disasters-human-response-nasa-mars-moon.html

"How Will Police Solve Murders on Mars?" from *The Atlantic*, by Geoff Manaugh, published September 14th, 2018, last accessed on November 28th, 2020, at https://www.theatlantic.com/science/archive/2018/09/mars-pd/569668/

The Intelligence Trap: Why Smart People Make Dumb Mistakes, by David Robson, W.W. Norton, 2019

"Learning on the Job: Studying Expertise in Residential Burglars Using Virtual Environments," in *Criminology*, by Claire Nee, Jean-Louis van Gelder, Marco Otte, Zarah Vernham, and Amy Meenaghan, volume 57, issue 3, pages 481–511, August 2019

"List of Sci-Fi Crimes That Will Become Possible by 2040: Future of Crime," from QuantumRun.com, published September 15th, 2020, last accessed on November 25th, 2020, at https://www.quantumrun.com/prediction/list-sci-fi-crimes-will-become-possible-2040-future-crime-p6

"Militarization, Measurement, and Murder in the High Arctic," in *Territory Beyond Terra* (Kimberley Peters, ed.), by Johanne Bruun and Philip Streinberg, Rowman & Littlefield, 2018

"A Multimillion-Dollar Criminal Crypto-Mining Ecosystem Has Been Uncovered," from *MIT Technology Review*, published March 25th, 2019, last accessed on November 24th, 2020, at technologyreview.com/s/613163/a-multi-million-dollar-criminal-crypto-mining-ecosystem-has-been-uncovered/

"Phantom of the ADAS: Phantom Attacks on Driver-Assistance Systems," from *The International Association for Cryptologic Research*, by Ben Nassi, Dudi Nassi, Raz Ben-Netanel, Yisroel Mirsky, Oleg Drokin, and Yuval Elovici, published January 28th, 2020, last accessed on November 28th, 2020, at https://eprint.iacr.org/2020/085.pdf

"Psychology in Deep Space" in *The Psychologist*, by Nick Kanas, volume 28, number 10, pages 804–807, October 2015

"The Self-Appointed Spies Who Use Google Earth to Sniff Out Nukes," in *The Atlantic*, by Amy Zegart, published December 6th, 2019, last accessed on November 28th, 2020, at https://www.theatlantic.com/ideas/archive/2019/12/new-nuclear-sleuths/602878/

"Someday, Someone Will Commit a Major Crime in Space," in *Slate*, by Jane C. Hu, published August 28th, 2019, last accessed on November 25th, 2020, at https://slate.com/technology/2019/08/space-crime-legal-system-international-space-station.html

"State Jurisdiction over Ice Island T-3: The Escamilla Case," in *Arctic*, by Donat Pharand, volume 24, issue 2, pages 81–152, June 1971

"True Crime: Murder on an Arctic Ice Floe," from *Mental Floss*, by Kara Kovalchik, published July 22nd, 2010, last accessed on November 28th, 2020, at https://www.mentalfloss.com/article/25261/true-crime-murder-arctic-ice-floe

"Vodka-Fueled Stabbing at Russian Antarctic Station: Here's What Psychologists Think Happened," in *Russia Today*, published November 2nd, 2018, last accessed on November 27th, 2020, at https://www.rt.com/news/442998-antarctic-stabbing-spoilers-vodka/

"What Life on Mars Will Be Like?" from *Slate*, by Taylor Mahlandt, published July 10th, 2019, last accessed on November 28th, 2020, at https://slate.com/technology/2019/07/robert-zubrin-mars-settlement-societies-community-government.html

"When It Comes to Living in Space, It's a Matter of Taste," from *Scientific American*, by Jim Romanoff, published March 10th, 2009, last accessed on November 28th, 2020, at https://www.scientificamerican.com/article/taste-changes-in-space/

"Why Deep-Learning AIs Are So Easy to Fool," in *Nature*, by Douglas Heaven, volume 574, issue 7777, pages 163-166, October 9th, 2010

INDEX

Note: Italic page numbers refer to illustrations.

abolition movement, 55–58
Aborigines, 25
Africa (slave ship), 34–35
AIDS/HIV, 169, 171–72
Alexander, Leo, 149n
alligators, 21–22
alternating current (AC), 96–97, 99–106, 108–11, 114, 120–21
American Medical Association, 150, 157
American Naturalist, The, 135
American Philosophical Society, 134
amphetamines, 232, 285
anatomical dissections
- of Charles Byrne, 69–72, 71n
- donation of bodies for, 82–83
- grave-robbing for, 9, 58–65, 67–72, 67–68n, 77, 79–80, 83
- and medical education, 62–64, 72, 77–78, 82–83
- and Webster-Parkman murder, 8, 84–95

Anatomy Act of 1832, 78, 81
anatomy riots, 78, 79–81
androids, 317
anesthesia, 90n
angiographs, 176–77
animals
- Harold Brown's experiments with, 96–97, 109–12, *111*, 114
- drug trials using, 118–19
- Thomas Edison's experiments with, 9, 95, 97, 106, 109–12, *111*, 112n, 114, 118, 119, 121
- John Hunter's experiments on, 117–18
- Nazi law on experiments, 146

Anne (queen of England), 30
Anning, Mary, 125
Antarctica, 309n
anti-genetics, 215–19
antipsychotic drugs, 196–98
anti-Semitism, 205, 226n
antisocial behavior, 246
Apatosaurus, 136n
Arctic ice islands, 307–9, *308*, 312
arsenic, 152–53
artificial intelligence, 300, 316–17, 320n
atomic bomb plans, theft of, 200–215, 219–24, 226, 226–27n, 228, 234, 304
Australia, 18, 22, 25–26, 257n, 314
AZT, 171–72

baking, in space, 310n
Ballou, William Hosea, 140–41

Banana Islands, 39, 48, 50
Bay of Campeche, 12–13, 16, 21
Beagle, voyage of, 32
Bentham, Jeremy, 82–83
Berlin, Andreas, 48–50
Bernoulli, Jacob, 299
Bethe, Hans, 226–27n
Bethlem Royal Hospital, *179*
bias, 44, 253, 256, 287, 291, 304
biological sex, and gender identity, 260–63, 261n, 269, 278–79, 282
biopiracy, 28–30
birds of paradise, 37–38n
Bitcoin, 314–15
Black, Tom, 203–5
Blade Runner (film), 317
blank slate theory of human nature, 219, 260–61, 264, 282, 302
Boston Latin School, 284
brain imaging, 176–77
brainwashing, 238
British Museum, 53
Brontosaurus, 136, 136n
Brothman, Abe, 220–21
Brown, Harold, 96–97, 109–14
buccaneers, 14–16, 14n, 21. *See also* piracy; privateers
Bunce Island, 35, 49
Burke, William, 59–61, 72–77, 91, 95
burking, 61, 72–73, *73*, 75–77
Byrne, Charles, 69–72, 71n
Byron, Lord, 233

Caesar's Last Breath (Kean), 90n
Canadian Broadcasting Corporation, 259
capital punishment, 106–8
Capone, Al, 233
Cassava, 48n
castrati, 270–71n
Catholic Inquisition, 24
Center for Disease Control, 158
Central Intelligence Agency (CIA), 199–200, 228, 230, 233–34, 236–38, 248, 253
chain-of-custody rules, 287, 295
Chicago World's Fair (1893), 120
chimpanzees, 119, 174–75, 178, 180, 181n, 249
China, 20, 30, 216, 234, 239–40
chlorpromazine, 196–98
chocolate, 52n
chromosomes, and gender identity, 260–62, 260n, 270
cinchona tree, 28–29
circumcision, 20, 254–55, 264, 270
Clarkson, Thomas, 58
Cleopatra, 3–5, 6
Cleveland, James, 39–40, 39–40n
Cleveland, William, 39–40n
climate change, 219
cocaine, 292, 296, 298
Cockran, Bourke, 114
Cody, William (Buffalo Bill), 130
Cold War, 200, 219, 228, 234, 236–38, 253
Coleridge, Samuel Taylor, 32
colonialism, 29, 32, 46
colonization of space, 10, 310–15
communism, 204, 208, 215
computers, and future crimes, 314–17, 320–21, 320n
Cook, James, 22, 48
Cope, Edward Drinker
  fossil excavation in New Jersey, 125–26, 126n
  fossil-hunting expedition in U.S. West, 132–35
  and Othniel Marsh's "salted" dig site, 122–23
  mining investments of, 139–40
  misinterpretation of *Elasmosaurus* skeleton, 127–28, 142
  rivalry with Othniel Marsh, 121, 123, 126, 128, 131–37, 140–44, 303
  and theory of evolution, 137
  and U.S. geological surveys, 139–42
  youth and education, 124–25
Cope, Julia, 133
copper, 100–101, 104
COVID-19, 172
criminality
  future computer crimes, 314–17, 320–21, 320n
  future genetic engineering crimes, 10, 300, 319–20
  and genius, 249–50
  in the name of science, 6–10
  space crime, 309–14

criminal justice, in space colonies, 312–14
cryptocurrencies, 314–15
crystal tests, 287
culture, and gender identity, 260–63, 267, 278–79, 282
Cushing, Henry, 71n
Custer, George Armstrong, 138
Cutler, John
  Guatemalan STD experiments by, 9, 159–60, 162–70, *167*, 169n, 170n, 302
  search for STD treatments for U.S. military, 160–62, *160*

Dalton, John, 299
Dampier, Judith, 15, 19
Dampier, William
  as captain of *St. George*, 31
  compared to John Hunter, 63
  expedition to Australia as captain of *Roebuck*, 22–27
  on foreign cultures, 25, 25–26n
  as naturalist, 12–16, *13*, 18–22, 19n, 24–28, 36
  navigation skills of, 11, 16–19, *17*, 21–22, 31
  as pirate, 12, *13*, 14–19, 21–23, 26–28, 30–32, 35, 303
  trial and conviction of, 11–12, 25–27
Darwin, Charles, 12–13, 24, 32, 137, 140
DeAngelo, Joseph James, 318
Defoe, Daniel, 31–32
democracy among pirates, 22–23, 23n
Diamond, Milton, 264, 277–79, 282
Dickens, Charles, 95n
digestion, 65n
dinosaurs, 123, 125–36, 126n, 136n, 143–44, 143n
direct current (DC), 96–97, 99–106, 108–10, 112, 114, 120–21
Disappearing Spoon, The (podcast), 29n
"Discourse on Winds" (Dampier), 21–22
diversity, 302
DNA
  and Golden State Killer, 318–19
  and Theodore Kaczynski, 246
  and plant evolution, 53
  and reviving extinct species, 319–20
Docherty, Margaret, 75
dogs, experiments on, 96, 106–7, 109–10, 112, 114, 119
Dookhan, Annie
  arrest and imprisonment of, 295–98, *297*
  deceptions about education of, 283–84
  and fraudulent drug testing, 284–90, *286*, 292–98
  as liar, 283–85, 298–99
Dr. Jekyll and Mr. Hyde (Stevenson), 65
drug identification testing, 284–90, 292–98
drug trials, 171–73
dry-labbing, 287–88, 295–96

eating, in space, 310–11, 310n
economies of scale, 315
ecstasy, 298
Edison, Thomas Alva
  and AC/DC competition, 97, 99–106, 108–12, 114, 120–21, 120n
  and electrocution executions, 107–8, 113–17, 121
  as inventor, 97–99, *98*, 102–3
  torture of animals by, 9, 95, 97, 106, 109–12, *111*, 112n, 114, 118, 119, 121
Einstein, Albert, 303–5
Elasmosaurus, 127–28, 142
electric chairs, 113–16, *115*, 121, 226
electricity
  AC/DC competition, 96–97, 99–106, 108–12, 114, 120–21, 120n
  torture of animals with, 9, 95–97, 106, 109–12, *111*, 112n, 114, 118–19, 121
electrocauterizing needles, 254
electrocution, as method of execution, 107–8, 113–17, 121
electroshock therapy, 178, 178n, 180
elephants, electrocution of, 112n
Elizabeth (slave ship), 54
encryption, 248n, 249
Equiano, Olaudah, 57–58
Escamilla, Mario, 307–9, 312–14
estrogen, 265, 271
ethics
  and Cleopatra's experiments, 3–5, 6
  and Annie Dookhan's fraudulent drug testing, 284, 287–90

ethics (*continued*)
and drug trials, 171–73
and future science, 300–305, 320
and Guatemalan STD experiments, 9, 163, 165–70, 166n
and intelligence, 303–5
and John Money's gender-reassignment surgeries, 265, 270, 279–80
motivations for unethical science, 7–9, 58
and piracy, 27
of Tuskegee syphilis study, 153–59, 157n, 163, 169–70
evolution, 53, 137, 219
executed criminals, anatomical dissections of, 62
executions of criminals, 106–8, 113–17, 121
executive function, 249
extinct species, revival of, 319–20

Falun Gong, 239–40
Federal Bureau of Investigation (FBI)
and forensic science, 292
and theft of U.S. atomic bomb plans, 202, 220–24, 226, 228
and Unabomber case, 247–48n, 248, 250–51, *251*
femininity, 256, 262–63, 266–67, 278, 282
fentanyl, 285
fetuses, experiments on, 4–5
Fisher, George, 23–27, 32
forensic anatomy, 84, 93–94
forensic science
drug identification testing, 284–90, 292–98
lack of scientific basis for, 290–92
and requirement for in-person testimony, 292–93
in space colonies, 312
formaldehyde, 285
Fothergill, John, 39
Franklin, Benjamin, 56
Freeman, Walter
attempts to defend legacy, 198–99
and development of lobotomy surgery, 173, 178n, 182–85, 302
and introduction of chlorpromazine, 196–98
and Egas Moniz, 177–78, 182–83, 195–96, 195n
on talk therapy, 178n
and transorbital lobotomies, 173, 187–96, *188*, 191n, *192*, 195n, 198–99
frontal lobes, 174–75, 180–82, 184, 187, 249
Fuchs, Klaus
arrest and confession of, 222–24
and Harry Gold, 201–2, 208–15, 219, 222–24
and theft of U.S. atomic bomb plans, 201–2, *203*, 209, 212–15, 226–27n, 228
Fuchs, Kristel, 211
Funes, Juan, 162–63

Galápagos Islands, 16, 32
Galileo, 299
gender identity
and biological sex, 260–63, 261n, 269, 278–79, 282
brain as biggest determinant of, 280n
and chromosomes, 260–62, 260n, 270
and culture, 260–63, 267, 278–79, 282
and genetics, 262, 269, 278
and genitalia, 256, 260–61
John Money's research on, 256, 260–62, 261n, 279–80, 302
gender-reassignment surgeries, 259–60, 259n, 263–72, 267n, 277n, 279–80. *See also* sex-reassignment surgeries
genetic engineering, future crimes involving, 10, 300, 319–20
genetics
bias against, 215–19
and breakdown of neurotransmitters, 246
and gender identity, 262, 269, 278
genetic testing, and criminal justice, 317–19
genitalia
ambiguous, 259–60, 280
fetal, 4
and gender identity, 256, 260–61

genus-species taxonomy, 38, 40
geographic "no man's lands," 307–9
George III (king of England), 69
Gibbins, Ernest, 27
Ginsburg, Ruth Bader, 293
Gold, Harry
  arrest and confession of, 222–26, *225*, 226n, 250
  and communism, 204, 208, 215
  disillusionment with spying, 206–7, 220
  and Klaus Fuchs, 201–2, 208–15, 219, 222–24
  in Lewisburg Penitentiary, 226–27
  motivations for spying, 204–5, 207–8, 303
  and theft of U.S. atomic bomb plans, 201–2, *203*, 209, 212–14, 219–26, 228
  youth during Great Depression, 203–4
Gold, Simon, 203
Golden State Killer, 318–19
gonorrhea, 9, 161–63, 166, 168, 302
Google Street View, 314
Grant, U. S., 139
grave-robbing
  for anatomical dissections, 9, 58–65, 67–72, 67–68n, 77, 79–80, 83
  and Abraham Lincoln, 81n
  sale of individual body parts, 83
  techniques of, 66–68, 67n, 68n
Gray, Ann, 75–76, 76n
Gray, James, 75–76, 76n
Greenglass, David, 213–14, 226, 228
Greenglass, Ruth, 226
Grey, Earl, 30
groupthink, 105, 302
Guatemala, John Cutler's STD studies in, 9, 159–60, 162–70, *167*, 169n, 170n, 302
Gulliver's Travels (Swift), 32

Hadrosaurus, 125–26
Haitian slave revolt, 58
Hall, Ted, 215
Halley, Edmond, 22, 45
Hamilton, Alexander, 80
hangings, 106
Hare, Margaret, 73
Hare, William, 59–61, 72–77, 91, 95
Harrison, Benjamin, 81
Harrison, John Scott, 81
Harrison, William Henry, 81
Harvard Medical School, 8, 84–87, 90n, 91–93, 95n, 290
Harvard Psychology Clinic, 233
Harvard University
  and anatomical dissections, 78
  Annie Dookhan's lies about attending, 283
  and Theodore Kaczynski, 229–30, 240, 242–44, 246
  and Henry Murray's abusive psychological experiments, 229–34, *232*, 238, 240
  and Webster-Parkman murder, 8
Haydn, Joseph, 65
Hayes, Rutherford B., 139
hermaphrodites, 256, 259, 264. *See also* intersex people
heroin, 285, 287, 296
Hess, Rudolph, 145
Himmler, Heinrich, 147
Hippocrates, 146
Hippocratic oath, 146, 150, 157
Hiroshima, Japan, 219
Hitler, Adolf, 233
HIV/AIDS, 119, 169, 171–72
Holmes, Oliver Wendell, Sr., 88
homosexuality, 27, 257n, 279
honesty, 300–304
hormones, 81–82, 256, 259n, 262–63, 270n, 277, 280
human nature, blank slate theory of, 219, 260–61, 264, 282, 302
human rights, and gender-reassignment surgeries, 280
human trafficking, 320, 320n
Hume, David, 65
humility, 291–92
Hunter, John, 63–65, *65*, 65n, 67–72, 74, 117
Hunter, William, 64
hurricanes, 12–14
Huxley, Thomas Henry, 137
hypothermia, Nazi experiments with, 147–49, 147n, *148*, 149n

immune systems, and weightlessness, 311
incest, 256
Indian Ring scandal, 138
informed consent, and malaria research, 171
inositol, 288
insane asylums
  and antipsychotic drugs, 196–97
  and John Cutler's STD experiments, 166–68
  and Walter Freeman's lobotomy experiments, 182–84, 193, 198, 302
  and Egas Moniz's lobotomy experiments, 178, 180–82
  psychiatric treatments in, 178–79, 178n, *179*, 238–39
insulin coma therapy, 178, 178n
intelligence, and ethics, 303–5
intelligence quotient (IQ), 31, 230, 241, 243, 249
International Space Station, 310n, 313
interrogation techniques, 228–35, *232*, 238, 240
intersex people, 256, 259–60, 262, 264, 280
Isles de Los, 33–34, 50

Jackson, Charles T., 90n
Jarisch-Herxheimer reaction, 153
Jay, John, 80
Jenner, Edward, 71
Joan of Arc phenomenon, 245
Johns Hopkins University, 256, 259, 265
Johnson, Virginia, 258
Joseph the Provider (Mann), 211
Jung, Carl, 231–33, 303

Kaczynski, David, 241, *243*, 247, 247–48n
Kaczynski, Linda, 247–48n
Kaczynski, Theodore
  arrest and imprisonment of, 251–53
  compared to brother David, 247, 247–48n
  as Harvard student, 229–30, 240, 242–44, 246
  manifesto of, 247, 247–48n, 250
  and Henry Murray's Harvard psychological experiments, 229–31, 238, 240, 244–46, 253
  as Unabomber, 230, 245, 247–53, 247–48n, 249n, *251*
  vulnerability to mental illness, 245–48
  youth and early education, 240–42, 242n, *243*, 246, 253
Kaczynski, Turk, 240–42, *243*, 248
Kaczynski, Wanda, 241–42, 244, 248
Kahneman, Daniel, 302
Kemmler, William, 113–17, 116n, 120–21
Kennedy, Joseph, 185–86, 186n
Kennedy, Kathleen, 186n
Kennedy, Rosemary, 185–86, 186n
ketamine, 298
Kinsey, Alfred, 258
Knox, Robert, 60–61, 74–75, *74*, 77
Kwasi (*lockoman*), 46–48, 48n

language acquisition, 261n
Lanning, Mary, 221
Leavitt, Donald "Porky," 307–8
Leidy, Joseph, 125–26, 126n, 136
lethal injection executions, 106–8
leucotomies, 181–83
Lewisburg Penitentiary, 226–27
lightbulbs, 97, 102–4
lightning, 105
Lightsy, Bennie, 308–9
limbic system, 180
Lincoln, Abraham, 81n
Linnaeus, Carl, 8, 38, 40, 44, 48, 52
Littlefield, Ephraim, 84, 87–90, *88*, 92–94, 95n
lobotomies
  chlorpromazine as chemical lobotomy, 196–98
  CIA's consideration of, 199–200
  compared to operations on transsexuals, 259
  Walter Freeman's development of surgery, 173, 178n, 182–85, 302
  Walter Freeman's transorbital lobotomies, 173, 187–96, *188*, 191n, *192*, 195n, 198–200
  and memory, 174–75
  Egas Moniz's experiments with, 178, 180–82, 184, 197

Longfellow, Henry Wadsworth, 91
lordosis, 263
Los Alamos, New Mexico, 210, 212, 215, 219, 226–27n
LSD, 240
Lysenko, Trofim, 215–19, 228

McKenty, Mary, 272
malaria, 27–28, 39, 54, 56, 170–71
malpractice, John Money's conduct of David Reimer's case as, 277, 277n
Mamani, Manuel Incra, 28–29
Manhattan Project, 208–9, 210, 219
Mann, Thomas, 211
MAOA gene, 246
marijuana, 19
Mars colonization, 10
Marsh, Othniel Charles
  And Cope's misinterpretation of *Elasmosaurus* skeleton, 127–28, 142
  and Brontosaurus identification, 136, 136n
  crusade for Indian rights, 137–39
  fossil-hunting expedition in U.S. West, 128–35, *129*
  and mounted uintathere skeleton, 126n
  rivalry with Edward Cope, 121, 123, 126, 128, 131–37, 140–44, 303
  salting a dig site for Edward Cope, 122–23
  and theory of evolution, 137
  and U.S. geological surveys, 139–42
  youth and education, 123–24
masculinity, 256, 262–63, 266–67, 275, 278, 282
Masters, William, 258
means-end fallacy, 9
medical education, and anatomical dissections, 62–64, 72, 77–78, 82–83
Melendez-Diaz, Luis, 292–94
Melville, Herman, 92, 232–33
memory, and lobotomies, 174–75
Mendel, Gregor, 299
Mengele, Josef, 147, 170, 173
mercury, 153
Merian, Maria, 43
meth, 298
mice, as experimental subjects, 118
Money, John
  bad temper of, 257
  as coiner of odd words, 260n
  conflict with Milton Diamond, 264, 277–79
  gender identity research by, 256, 260–62, 261n, 279–80, 302
  and gender-reassignment surgery, 259–60, 259n, 263–72, 267n, 277n, 279–80
  provocative stands on social issues, 256–57, *258*
  and David Reimer, 267–72, 277n, 278–80, 302
  sex life of, 258–59
  youth and education, 255–56
Moniz, Egas
  angiograph research of, 176–77
  on chimpanzee experiments, 174–75, 180, 181n
  and Walter Freeman, 177–78, 182–83, 184, 195–96, 195n
  legacy of, 198
  lobotomy experiments, 178, 180–82, 184, 197
  as neurologist, 175–76, 183
  portrait of, *176*
  on talk therapy, 178n, 181n
Montgolfier brothers, 56
Morgan, Christiana, 231
morphine, 285
Morton, William, 90n
Mosquirix, 170–71
motivations for unethical science, 7–9, 58
Murray, Henry
  abusive psychological experiments by, 229–34, *232*, 238, 240, 244–47, 253
  and Carl Jung, 231–33
  and John Money, 256
  World War II OSS work, 233–34
Murray, Josephine, 231–32
Murray, Thomas, 13

Nagasaki, Japan, 219
National Academy of Sciences, 139
Natural History Museum of London, 53
Nazi World War II experiments, 10, 145–51, 147n, *148*, 149n, 159, 173

Neanderthals, 319–20
Nelson, Horatio, 22
neurosis, and lobotomies, 175
neurotransmitters, 197, 246
Newton, Isaac, 8, 22, 45, 52, 53, 299
New Voyage Round the World, A (Dampier), 19, 21
nihilism, 249–51
Nobel Prize, 177, 196, 302
nuclear weapon plans, theft of, 200–215, 219–24, 226, 226–27n, 228, 234, 304
Nuremberg Code, 150–51, 244

Oberlin College, 242
octopuses, 119, 119n
Office of Strategic Services (OSS), 233–34
On the Origin of Species (Darwin), 24
opiates, 285
Orwell, George, 262
Outer Space Treaty of 1967, 309
Oxford English Dictionary, 20

Palin, Sarah, 219
Parkinson, James, 71
Parkman, George, 8, 84–94, *84*, *91*, *94*, 95n, 290
Parks, Rosa, 159
Pavlov, Ivan, 261
Pawnee people, 129–30
Peabody, George, 123, 126, 128
pedophilia, 257, 257n
penicillin, 153–54, 157n, 163
penises
  and gender-reassignment surgeries, 266–67
  John Money on, 255
  and David Reimer's circumcision and gender reassignment surgeries, 255, 274–76
Penn Sugar, 203–5, 204n, 212, 220
Pepys, Samuel, 21
personality testing, 23–234
phalloplasty, 276–77
phimosis, 254–55
piracy. *See also* buccaneers; privateers
  biopiracy, 28–30
  of William Dampier, 12, *13*, 14–19, 21–23, 26–28, 30–32, 35, 303
  democracy among pirates, 22–23, 23n
  and ethics, 27
  peak times of, 14n
Pitt, William, 65
Playboy, 258
Plutarch, 3
poisons, Cleopatra's use of, 3–4
political dissidents, psychiatric treatments for, 238–40
poverty, and anatomical dissection, 77–83
praise, 285, 289
premortems, 302
Principia Mathematica (Newton), 45, 52
prisoners, medical experiments on, 3–4, 10, 146–47, *148*, 161–63, 173, 227–28
privateers, 14–15, 23, 30, 31, 55. *See also* buccaneers; piracy
problem-solving, and lobotomies, 174–75
prostitution, 162–64
protozoans, 170
psychiatric treatments
  and gender-reassignment surgeries, 270, 280
  in insane asylums, 178–79, 178n, *179*, 238–39
  and political dissidence, 238–40
  for transsexual people, 259, 280
psychoanalysis, 232
psychological torture, Theodore Kaczynski as experimental subject, 229–31, 238, 244–45
psychosexual engineering, 264
psychosurgery, 180–83, 181n, 187, 190, 195–96, 302. *See also* lobotomies
pterodactyl, 131, 133, 144
puberty, 270–71, 270–71n
Public Health Service
  and John Cutler's STD experiments, 160, 162–69
  Tuskegee syphilis study, 151–59, *153*, 157n
Putin, Vladimir, 218

Quassia amara, 48
quinine, 28, 29n, 46, 56

Red Cloud, *129*, 138–39
Reimer, Anthony, *276*
Reimer, Brian, 254, 266–69, 274–75, 281
Reimer, David (née Bruce, née Brenda)
botched circumcision resulting in loss of penis, 254–55, 264–65, 270
as Brenda, 266–74, 276, 278, 280–81
conversion from Brenda to David, 273–75
gender-reassignment surgery, 265–66, 274–75, 277n
marriage to Jane, 276, *276*, 281
and puberty, 270–71, 270–71n
relationship with John Money, 267–72, 277n, 278–80, 302
suicide and suicidal thoughts of, 269, 275, 281–82
Reimer, Jane, 276, *276*, 281
Reimer, Janet
affairs and suicide attempts, 269–70
difficulties raising David Reimer as girl, 266–73
and John Money, 260, 265, 270
and David Reimer's botched circumcision, 255, 265, 272–73
and David Reimer's gender-reassignment surgery, 265–67, 267n
and David Reimer's suicide attempts, 275
Reimer, Ron
alcoholism of, 269
difficulties raising David Reimer as girl, 266–73
and John Money, 260, 265
and David Reimer's botched circumcision, 255, 265, 272–73
and David Reimer's gender-reassignment surgery, 267, 273
and David Reimer's suicide attempts, 275
resurrectionists
as confidence men, 67, 84
grave-robbing of, 9, 65–66, 68
illustration of, *63*
and Ephraim Littlefield, 88, 92
punishment of, 68, 68n
and sale of body parts, 83
Reverby, Susan, 169, 169n, 170n
"Rime of the Ancient Mariner, The" (Coleridge), 32
Rivers, Eunice, 155–59, *156*, 157n
Robinson Crusoe (Defoe), 31
robots, 10, 18, 317, 325
Roebuck (ship), 22–27
Rosenberg, Ethel, 202, 214, 226, 228
Rosenberg, Julius, 202, 214, 226, 228
Royal Society, 21–22, 39, 45, 53, 55, 58, 65
rubber, 30

St. George (ship), 31
scale calibration, 287
Scalia, Antonin, 293
science. *See also* ethics
criminality in the name of, 6–10
in forensics, 290–92
and piracy, 27
slavery's influence on, 8, 10, 32, 34, 36–38, 43, 45–46, 52, 54
trustworthiness of, 298–99, 304–5
scientists. *See also specific scientists*
madness of, 4–5
obsession-driven misdeeds of, 6, 10
self-confidence of, 298–99
Selkirk, Alexander, 31
sex discrimination, 282
sex-reassignment surgeries, 259–60, 259n, 263–72, 267n. *See also* gender-reassignment surgeries
sex robots, 317
sexual differentiation, 4
sexual identity, John Money's research on, 256, 260–62, 261n
sexually transmitted diseases, 9
Shakespeare, William, 72
Shapley, Harlow, 95n
Shaw, Robert, 86
ship surgeons, 38, 38n
Sierra Leone, 33, 36, 39, 44, 48–49, 51, 56
Simpson, Abigail, 73
Sixth Amendment, 292–93
slave trade
and abolition movement, 55–58
influence on science, 8, 10, 32, 34, 36–38, 43, 45–46, 52, 54

slave trade (*continued*)
and modern technology, 320, 320n
in Sierra Leone, 34–36, 44–45, 49–52, 56
slave ships, 34–35, 37–38, 49–51, 54
and Henry Smeathman, 34–36, 38–39, 44–45, *47*, 49–52, 54–56
Sloane, Hans, 52–53, 52n
smart technologies, 315–16
Smeathman, Henry
marriage to Brunetta, 40
as naturalist, 33, 35–36, 39–44, *47*, 48–49, 51, 54–56
relationship with African guides, 43–44
Sierra Leone expedition, 33–36, 39–45, *47*, 48–52
and slave trade, 34–36, 38–39, 44–45, *47*, 49–52, 54–56
study of termites and mounds, 40–44, *41*, *42*, 55, 64
Smith, Adam, 65
Smith, Lemuel, 107
Smithsonian Institution, 125, 140
Southwick, Alfred, 107, 116
Soviet Union
fear of the Red Menace, 228
and Trofim Lysenko's anti-genetics views, 215–18
and theft of U.S. atomic bomb plans, 204–15, 221–22
space exploration, 300, 309–15, 310n
Sparrman, Anders, 57–58
Stalin, Joseph, 216–19, 234
STDs, John Cutler's STD experiments, 9, 159–70, *167*, 170n, 302
stem cells, 319
Stevenson, Robert Louis, 65
stress, and ethical lapses, 289, 294
subconscious mind, 233
sub-species, 24
sudden infant death syndrome (SIDS), 82
sulfuric acid, 285
Swift, Jonathan, 32
syphilis
Guatemalan STD studies, 9, 163, 165–68, 166n, 302
penicillin treatment for, 153–54, 157n, 163
Tuskegee study of, 151–59, *153*, 157n, 169–70
Systema Naturae (Linnaeus), 38, 52

talk therapy, 178n
termites and mounds, 40–43, *41*, *42*, 42n, 55, 64
Tesla, Nikola, 101–4, *102*, 120n, 121
testicles, 275
testosterone, 270–71n, 273–74
thermal diffusion, 210
Thomas, Clarence, 293
thymus gland, 82
Time, 269
Tittle, John, 34–35, 50
torture
of animals, 9, 95–97, 106, 109–12, *111*, 112n, 114, 118–19, 121, 146
psychological, 229–31, 238, 244–45
transgender people, 259n, 269, 279–80. *See also* gender identity; transsexual people
transorbital lobotomies, 173, 187–96, *188*, 191n, *192*, 195n, 198–99
transsexual people, 259–60, 259n, 271, 277n, 279–80. *See also* gender identity; transgender people
Treehorn, Jackie, 280n
triangular trade, 37, 51, 55
Trump, Donald, 120n
Tuskegee syphilis study, 151–59, *153*, 157n, 163, 169–70
Twain, Mark, 95n

Uganda, 27, 171–72
Unabomber. *See* Kaczynski, Theodore
unclaimed corpses, for dissection, 77–78, 81–82
underground electric utilities, 100, 105–6
United Fruit Company, 162
United Nations, 280
U.S. geological surveys, 139–42
U.S. National Academy of Sciences, 290
University of Kansas, 263
unjust convictions, and fraudulent drug testing, 288–89, 296–97

uranium, 210, 215
urethras, 161, 164, 168, 275
urination, 266, 272

vaccines, 170–71, 227, 283–84, 289
vaginas, and gender-reassignment surgeries, 265
Voltaire, 117
von Steuben, Baron, 80

Wadström, Carl, 57–58
Wallace, Alfred Russel, 39, 39n
Warren, John, 90n
Washington, Booker T., 159
Washington Post, 158, 250, 277
Watts, James, 183–84, 189–90, 195
Webster, John White, 8, 84–95, *85*, 90n, *94*, 95n, 243, 290
Westheimer, Ruth, 258
Westinghouse, George, 103–6, 108–9, 112–14, 117, 120–21, 120n
Wickham, Henry, 30
woolly mammoths, 319
World Health Organization (WHO), 170–71
World War II
  Henry Murray's OSS work, 233–34
  Nazi experiments during, 10, 145–51, 147n, *148*, 149n, 159, 173
  Nazi-supplied quinine during, 29n
  and nuclear weapons, 210, 219, 304
  STDs in military during, 160
writing, in 17th century, 19n

X-rays, 176–77
X/X chromosomes, 261
X/Y chromosomes, 260, 261

Yale University, 123, 126–28, 131, 134, 138, 174–75
Yolngu people, 257n
Young, Brigham, 130–31

Zedong, Mao, 234

ABOUT THE AUTHOR

SAM KEAN is the *New York Times* bestselling author of *The Bastard Brigade*, *Caesar's Last Breath* (the *Guardian*'s Science Book of the Year), *The Tale of the Dueling Neurosurgeons*, *The Violinist's Thumb*, and *The Disappearing Spoon*. He is also a two-time finalist for the PEN / E. O. Wilson Literary Science Writing Award. His work has appeared in *The Best American Science and Nature Writing*, *The New Yorker*, *The Atlantic*, and the *New York Times Magazine*, among other publications, and he has been featured on NPR's *Radiolab*, *All Things Considered*, and *Fresh Air*. His podcast, *The Disappearing Spoon*, debuted at #1 on the iTunes science charts. Kean lives in Washington, D.C.